T0074427

PERIPHERY

PERIPHERY

HOW YOUR NERVOUS SYSTEM PREDICTS AND PROTECTS AGAINST DISEASE

MOSES V. CHAO

HARVARD UNIVERSITY PRESS
Cambridge, Massachusetts
London, England
2023

Printed in the United States of America

First printing

Library of Congress Cataloging-in-Publication Data

Names: Chao, Moses V., author.
Title: Periphery : how your nervous system predicts and protects against disease / Moses V. Chao.
Description: Cambridge, Massachusetts ; London, England : Harvard University Press, 2023. | Includes bibliographical references and index.
Identifiers: LCCN 2022062082 | ISBN 9780674972308 (cloth)
Subjects: LCSH: Nerves, Peripheral. | Health risk assessment. | Disease susceptibility.
Classification: LCC QP365.5 .C436 2023 | DDC 612.8/1—dc23/eng/20230428
LC record available at https://lccn.loc.gov/2022062082

To Lincoln and Beatrice

CONTENTS

ABBREVIATIONS

ALS	amyotrophic lateral sclerosis
ASD	autism spectrum disorder
BDNF	brain-derived neurotrophic factor
CIPA	congenital insensitivity to pain
CMT	Charcot-Marie-Tooth disease
CNS	central nervous system
CSF	cerebrospinal fluid
CTE	chronic traumatic encephalopathy
DRG	dorsal root ganglia
ELP1	elongator protein 1 (also known as IKBKAP gene)
GI	gastrointestinal (tract)
HSANs	hereditary sensory and autonomic neuropathy disorders
IBS	irritable bowel syndrome
IGF-1	insulin/insulin-like growth factor
NGF	nerve growth factor
PNS	peripheral nervous system
PTSD	post-traumatic stress disorder

SMA	spinal muscular atrophy
TrkA	tropomyosin receptor kinase A
TRP	transient receptor potential

INTRODUCTION

> The scientist is motivated primarily by curiosity and a desire for truth.
>
> —IRVING LANGMUIR

THE WORD *PERIPHERAL* usually refers to a remote outlier, an object on the boundary or fringe. But in the context of this book, the peripheral nervous system (PNS), or "periphery," is in no way an outlier. It is front and center, a vital part of the body's everyday functioning that, as we are now learning, plays a key role in its dysfunction.

The periphery is an elaborate set of nerve networks that exists outside the brain and spinal cord, the central nervous system (CNS). Functioning similarly to how it did millions of years ago, our periphery detects sensations—smells, sounds, and touch—and transfers information about the environment to the rest of the body. Beyond providing immediate feedback, the system also controls many organs that regulate digestion and heart rate.

But if the peripheral nervous system is such an integral part of our body's health, why isn't it as well known as the CNS? The answer to this question may lie in part in the overwhelming emphasis, in terms of funding and research, on the brain and the CNS. Only in the past few years has research revealed that we need to explore the role of the periphery in more detail, especially in relation to health and disease. I believe that if we consider our body's health and its potential for disease without knowledge of or appreciation for the peripheral nervous system, it would be like listening to an orchestra without a string section—the very foundation of the ensemble's sound would be missing.

We can define the periphery most generally as being outside the central brain and spinal cord. But this demarcation is an oversimplification and very arbitrary; many networks of nerves in the brain connect with peripheral nerves, and vice versa, and the peripheral nervous system connects the organs with the brain. Keep this overlap in mind as you migrate through the chapters. The peripheral nervous system directly interacts with the CNS.

In the pages that follow, I argue that the periphery is a unique and under-recognized yet vital part of our body that offers important clues predictive of the onset of disorders and diseases. I highlight the periphery's critical role in the body, from controlling organs that regulate digestion, sweating, touch, and heart rate to maintaining the body's overall health. When there is a malfunction in the body, symptoms are often directly connected to the periphery. In this way, the periphery offers early warnings of future disease.

You will see that disturbances of the peripheral system are associated with numerous diseases, although not all of these associations are considered to be entirely proven. Reading this book and contemplating its ideas will require a shift in perspective and an openness of mind, both of which are imperative in scientific and medical research. Researchers know well that not every hypothesis will be proven correct. Yet they also know that scientific progress depends on exploring new theories based on observation and analysis. Some discoveries take place after an unanticipated result or a serendipitous event; some are developed over time, based on observation, thoughtful data collection, and contemplation. This book takes the latter approach in an effort to progress toward a greater understanding of the peripheral nervous system's critical role in health and disease.

Although there may be numerous neurological and psychiatric diseases that have associated events or direct connections in the periphery, my focus in this volume is on a handful of disorders to make the case: Parkinson's disease, Charcot-Marie-Tooth disease, Riley-Day syndrome, and autism. Parkinson's disease exhibits symptoms that emanate from the peripheral nervous system, many of which occur well in advance of diagnosis or later-stage disease.[1] In Charcot-Marie-Tooth disease, a group of inherited disorders that cause peripheral nerve damage, effected nerves are unable to send signals to other nerves, resulting in difficulty with motor functioning in the lower extremities. Riley-Day syndrome, or familial dysautonomia, is genetic and affects the development and survival of nerve cells in the autonomic nervous system, leading to difficulties with swallowing reflexes, body temperature, and muscle tone.[2] While most explanations for autism, especially in children, focus on brain size and circuitry (autism's connection to the periphery is not widely accepted), recent findings on the sense of touch and the role of the senses more broadly show the peripheral nervous system to be an important factor in this complex disorder and support the need for further research.

The final chapters of this book build on this knowledge of the periphery's connections with disease by also considering its plasticity, or capacity for change. Malleability to injury and the ability to alter its environment open profound opportunities for the role of the periphery in fighting disease, and medical researchers are looking to harness this capability, primarily for therapeutics. Through discussion of recent findings and potential targets for future efforts, I consider how peripheral problems could possibly be reversed.

Throughout this book, I explore the periphery's largely unrecognized and untapped potential for health and healing. But don't just take my word for it. At the annual meeting of the Society for Neuroscience in 2015, Francis Collins, the director of the National Institutes of Health (NIH), gave a standing-room-only talk about progress in the field of neuroscience. At the end of his speech, Dr. Collins stressed the need to focus on the peripheral nervous system. For the director of NIH to encourage more studies on the periphery was overdue recognition and a sign of its potential. His speech is one of the many reasons I chose to write this book.

As we will see in the course of this book, the NIH's recent acknowledgment of the periphery's ability and potential comes more than a century after the critical observations and discoveries of a handful of outstanding scientists and physicians, including the now-famous physician James Parkinson, who in the nineteenth century observed and recorded warning signals from the periphery. I was, and continue to be, greatly impressed with the original and groundbreaking observations that were made by Parkinson and others and their boundless curiosity, intuition, and instinct that led to discovery.

But before I tell the story of the scientists, their ideas, and the seminal findings that have led us to what we know today, we must start at the very beginning. To do so, we journey to the dark origins of the peripheral nervous system—before humans existed.

1

THE FIRST SYSTEM

> The uniformity of the earth's life, more astonishing than its diversity, is accountable by the high probability that we derived, originally, from some single cell.
>
> —LEWIS THOMAS

IN THE BEGINNING, THERE WAS A WORM. An anatomically simple—but completely functional—organism that lived in a deep-sea bed, this common ancestor of vertebrates and invertebrates had a nerve net, a mesh-like system of nerve cells dispersed across the body that allowed it to detect and respond to light and touch. This ancestral organism thrived in its environment, and eventually it reproduced. After generations and over the course of evolution, one of the worms turned upside down, resulting in a centralized central nervous system on the backside of its body.

This evolution story was first told by the German scientist Anton Dohrn (1840–1909). Dohrn proposed that vertebrates—including humans—originated from a simple organism, a primitive annelid worm, nearly six hundred million years ago. According to this now established theory, the peripheral nervous system thus evolved first, before the appearance of a brain and the rest of the central nervous system.

Five hundred million years later, in 2012, technological advancements have allowed us to consider our evolution—and Dohrn's evolutionary story—using genetic analysis. Researchers at Stanford University found similarities between the early acorn worms and our vertebrate brains in molecular and genetic blueprints.[1] This 2012 data supported work from Swedish scientists showing that the early embryonic development of the worm *Xenoturbella bocki*, which had a nerve net but lacked a brain, was similar to that of humans.[2] Both studies support Anton Dohrn's theory from the 1860s: we evolved from a worm that had a peripheral nerve net, not a brain.

For millions of years, the peripheral nervous system has been a conveyer of information. It is the first system in our body to detect sensations like smells, sounds, and touch. Information about the environment—heat, cold, and pain—must be detected quickly, and it is the periphery's job to transfer information about one's surroundings to the rest of the body. Whether for a vertebrate in the ocean—"Danger! Predator!"—or a human—"Pull your hand away from the hot stove!"—sensory data is vital for cognitive and physical response.

Today, our human bodies have evolved the nerve net into an entire system, which includes the sensory nerves and ganglia (sensory fibers and neuron cell bodies), as well as the heart, kidneys, gut, and other organs that function with close ties to the nerves and that regulate digestion and heart rate. Consider the periphery's elaborate set of nerve networks, spinal roots, and sensory and autonomic ganglia as a partner to the central nervous system (the brain and spinal cord); the PNS and the CNS together make up the body's nervous system. Although we are just now beginning to grasp how the periphery is responsible for all the vital connections and wiring outside the CNS and how it affects blood flow, heart rate, oxygen exchange in the lungs, and even the contraction of the eye muscles, it is clear that the periphery plays an important part in communication. Far from being "outside," the PNS is a defining part of the body, like the string section of an orchestra—front and center, vital to the sound and function of the ensemble.

The String Section

To truly understand the function of the peripheral nervous system, we must explore its parts. For this purpose, I like to use the string section of an orchestra as an analogy. A string section is made up of violas (the sensory nervous system), violins (the autonomic nervous system), cellos, and string basses. Perhaps counter to what you would expect, we'll begin with the violas.

The violas play a key role in most orchestral music, often producing long harmonic phrases that bridge the violin's melody line with the bass notes of the cello and string bass. To play their complementary parts with full resonance, the viola players must be exceptionally aware of the other sections. Similarly, the sensory nervous system must detect a change in the environment. To do so, the sensory nerves (violas) are a part of the dorsal root ganglia (DRG) and several cranial ganglia, which extend away from the brain and spinal cord. The DRG comprises numerous sensory neurons, cell types of different sizes, electrical properties, and gene expression that allow us to feel sensations ranging from pain to pressure and temperature.[3] When the neurons sense pain or a blast of heat and you move your hand away from the stove, the communication travels quickly to inform the brain and the rest of the CNS.

With their bright tone, violins usually play the melody line. In orchestral pieces, they are grouped into firsts, seconds, and sometimes thirds to add texture and harmony. Similarly, in 1921, English physiologist John Langley proposed that the autonomic nervous system was best understood in three divisions, a structure we still use today: the sympathetic, parasympathetic ("beside sympathetic"), and enteric systems.[4] We will consider these the first, second, and third violin sections—all part of the violin section yet playing different, though complementary, parts. Like musicians, the parts of the autonomic nervous system are closely aligned, must work together, and must switch quickly from activity to inactivity. Responsible for control of bodily functions that occur involuntarily, like our heartbeat and digestion, the autonomic nerves are sophisticated sensors capable of detecting subtle changes in the body that keep the internal organs in proper working order; the autonomic nervous system acts as a thermostat for the body. Sometimes referred to as the "move and drool" system, the autonomic system controls all the muscles and glands.

The sympathetic and parasympathetic networks (violins I and II) each comprise large sets of nerve cells and have opposing functions, serving as a balance for each other. The sympathetic system stimulates the body by speeding the heart rate and increasing blood flow, whereas the parasympathetic system frequently slows things down, allowing the body to rest by decreasing the heart rate. The sympathetic nerves form a ganglia—a collection of neuronal bodies—just outside the spinal cord like a long necklace of pearls, connecting internal organs. The parasympathetic nerves are in ganglia located in the brainstem, in the sacral spinal cord, and in close proximity to the organs they regulate.

The third division, the enteric nervous system, has a name derived from the Greek word *enterikós*, or intestines. Like the third violin's part, it plays so well in harmony with the other sections that it sometimes goes unnoticed; it is even often omitted from many textbooks.[5] However, with its role as the sole controller of the digestive system, the enteric system is critical to our body's health. It senses changes and regulates secretions in the gut through sensory and motor neurons that communicate directly with the gut and the musculature, independent of what the brain is doing. Our growing realization of the enteric nervous system's role in the gut has led researchers and scientists to a new understanding of our evolution and the potential in our body for healing.

When we are born, our guts are colonized with an astounding number of different species of microbes (more than one hundred trillion bacteria belonging to close to a thousand species) weighing two to four pounds and filling two to four quarts. In contrast to the twenty thousand genes in the human genome, there are more than three million genes spread out across the microbiome, including the collection of microorganisms in the gut.[6] Remarkably, although humans share many of the same gut bacteria, approximately two-thirds of an individual's bacteria are unique, and each of these species of bacteria engages in communication with many organs, including the brain.[7] Bacteria have an enormous say about what goes on in the body.

During infancy, we learn to smell, taste, and eat, and we experience the world through our abilities to hear, touch, and taste and, eventually, through coordinated movement. In those early days, there is a pathway—an open sensory gateway—from the gut or digestive system to the brain. Although this pathway remains open as we age, it is believed that its presence in the early stages of development has a real effect on communication. The brain, spinal cord, peripheral and nervous system, and the digestive, endocrine, and immune systems sense and communicate with each other through different cell types. The digestive system gathers information and responds to food, chemicals, temperature fluctuation, and our environment like a sensor and communicates that information to the brain. The gut-brain connection also works to monitor the regulation of body functions, such as the protection of microorganisms and digestion of foods.

Most people are familiar with one of the gut's main roles: breaking down and absorbing nutrients from the food we eat. Yet the gut—the gastrointestinal system—has other critical responsibilities that support body functioning, including within the immune system. Even as the gut absorbs nutrients and calculates the chemical composition and the calories of ingested food, different physiological states of the gut wall are detected not only by the gut microbiota but also by enteric nerves.[8]

We now understand that the spider-web-like net of sensory neurons that covers and runs along the digestive tract in close contact with muscles and blood vessels is critical to our body's health. Containing more than one hundred million neurons, the enteric system surrounds the digestive system and acts as the control panel for gastrointestinal events, orchestrating the movement of food through the entire gastrointestinal tract, which measures eight to ten meters in length, by controlling the muscles surrounding the gut. To process the tons of food consumed during a lifetime while being exposed to an assortment of toxic materials, the network of enteric nerves that surrounds the gut is just as large and complex as the brain and uses the same neurotransmitters and signals. However, while it is a part of the peripheral nervous system, the enteric system is also an independent entity and has its own set of guidelines.[9] The enteric nerves largely operate separately from the brain and spinal cord, working instead with sympathetic and parasympathetic parts to regulate gut functions. And enteric neurons and neighboring cells, including immune cells, serve a protective role in the gut, acting as a gatekeeper and controller of the bacteria populations.

Superimposed on the enteric nerves and intestines, immune cells are more numerous in the gut than anywhere else in the body. The gut also has remarkable defenses against foreign invaders, including a muscular barrier that acts as a defensive wall to prevent harmful chemicals and microorganisms from entering. Further, a layer of cells lines the gut and serves as a protective boundary between the gut and the blood circulation of the host. And in the large intestine, there are two layers that serve as physical barriers between gut bacteria and immune cells, although even with all these boundaries, small numbers of bacteria can still move through intestine walls. Consider it an enormous battlefield—armies of immune cells and battalions of neurons in continual communication with the bacteria in the gut to fight foreign invaders and to determine future decisions and the course of the enteric system.[10]

The GI tract is the second-largest sensory organ of the body (second only to the skin), and it monitors the environment incessantly. Embedded in the gut wall are a vast array of receptors that are key to the gut's sensory perception. The same receptors are also found in our taste buds in the tongue, where they recognize specific sensations in food, such as sour and sweet, but since they are located in the gut—in the periphery—these receptors can detect many stimulants in the form of sensory information throughout the body. So the connectivity between stomach and brain, as well as between the enteric nervous system and immune system, is important to understanding brain-gut communication.[11]

Michael Gershon, an acclaimed professor at Columbia University, pioneered studies of the enteric nervous system and frequently refers to the gut as having a mind of its own, a "brain in the bowel."[12] One could say that the enteric system not only acts as the third-chair violins but also operates as its own ensemble, with a different reach and audience.

But regardless of whether we consider the enteric system a separate orchestra or part of the string section in the periphery, each division of the nervous system must interact with the multiple components. The peripheral nervous system functions not in isolation but in continuous interaction with the brain and spinal cord. Specifically, the sympathetic and parasympathetic nerves extend to the skin and all the internal organs, small muscles, blood vessels, and glands, and the enteric nerves handle the digestive tract. So although our experiences are ultimately processed by the brain, the periphery is adept at receiving, saving, and disseminating information, and there is an active, continuous interplay between the periphery and the CNS, with considerable direction from internal organs like the gut and the immune and endocrine systems. This crosstalk is a sharing of motifs.

Crosstalk

The peripheral nervous system detects and communicates sensations with speed and precision, because even the smallest misstep can impact normal reactions, and repercussions can develop into maladies with far-reaching effects. Once a sensation is detected, the periphery sends warnings of an impending injury, ailment, or illness. The CNS responds—"Pull that hand away"—and the endocrine system reacts, signaling to the pituitary gland to produce a hormone that tells the adrenal glands to increase the production of cortisol. Released into the bloodstream, the stress hormone cortisol raises glucose levels and stimulates reactions in the gut and heart. This is the stress, or fight-or-flight, response; the body is ready to react further, as necessary. The sympathetic and parasympathetic systems balance each other much like the first and second violin parts.

To do so, sympathetic nerves stimulate to increase blood flow and heart rate during or after stressful situations. In turn, the parasympathetic system generally does the opposite, dampening responses and lowering heart rate. The two systems work together to fine-tune and adjust their actions with the enteric system. Similar to a yin-and-yang mechanism that represents opposite duties, the sympathetic and parasympathetic systems act in a complementary manner to monitor the rich sensory environment the body encounters. The capacity for such balance within the body is imperative. The push/pull system of autonomic nerves allows us to conserve energy and nutrients but also to trigger resources when there is a need for a more active energetic state; the autonomic system ensures that our cardiac, pulmonary, reproductive, and digestive systems are in sync.

The periphery also interfaces with the immune system, as continual production of cortisol or too many stressful events can affect health and lead to increased inflammation or a weakened immune system. As a watchful surveillance system, the periphery offers many clues to understand the mechanisms behind normal and abnormal states. And beyond our physical health and safety, our senses are related to our cognition, attention, and perception, even our memory; they are a part of how we make sense of the world around us. Remarkably, the peripheral nervous system does all of this without our help or our control, and often without our awareness. However, our lack of attention can be problematic at times.

Signals or clues from the periphery can precede a disorder or disease by years, even decades. These small incidents modulate sensory (afferent) and motor (efferent) connections. Signals from the brain to the spinal cord reach the periphery by efferent (from the Latin, "to bring away from") motor neurons to control breathing and digestion. The peripheral nervous system reports back to the brain by relaying information through afferent ("to bring toward") sensory nerves. A breakdown of this communication from the peripheral nervous system, or our failure to pay attention to warnings, has an impact on the gut, heart, lungs, liver, and kidneys, as well as on sensory and motor autonomic nerve fibers that control the networks.

These clues are often symptoms that one may have experienced before, including constipation, pain, sweating, and a racing heart, all of which are emblematic of aberrations in the peripheral nervous system. An apt quote attributed to Muhammad Ali, "It isn't the mountains ahead to climb that wear you out; it's the pebbles in your shoe," suggests the significance of the little things we ignore or simply fail to notice. While they may lead to different diagnoses or larger problems in the body, the symptoms are clues from the periphery that communicate that there is a malfunction or a disorder. They can be predictors of other disease states or early indications of things to come.

2

A GUT REACTION

> Scientists are cautiously beginning to question the view that the brain is the sole and absolute ruler over the body. The gut not only possesses an unimaginable number of nerves, those nerves are unimaginably different from the rest of the body. The gut commands an entire fleet of signaling substances, nerve-insulation materials, and ways of connecting.
>
> —GIULIA ENDERS

BRAIN DISEASES FIRST FORM IN THE GUT.[1] Although one could argue that this idea and much of what is presented in this chapter is speculative, it should also be recognized that the ability to theorize about the implications of experimental data is an essential part of the scientific process. We must consider ideas that arise from testing hypotheses and suggesting new experiments. Further, there is quite a bit of research to support the idea that the gut—a major part of the periphery—plays an important role in pain, disorders, and diseases. In fact, the theory that many conditions—including brain diseases, neurological disorders, pain, and metabolic conditions such as obesity—first form in the gut has recently garnered much interest from researchers.

Recall from the previous chapter that the guts of newborns are colonized with an astonishing number of microbes and that the GI tract is the second-largest sensory organ of the body, second only to the skin in size. The sensory nerves are incessantly monitoring the environment, including the bacteria within the gut. As a matter of fact, current speculation is that certain maladies, including constipation, are due to the gut bacteria.

Since the microbial environment is modulated by autonomic control of movement of food through the intestines, any changes in the GI tract alter the timing of contractions and, accordingly, the transit times of food. Shifts in the numbers of bacteria result in reduced contractions in the intestines and lead to constipation. We can also consider the slowing of defecation as a breakdown of communication between the peripheral nerves and the digestive system. Microbiota genes allow bacteria to produce neurotransmitters, such as dopamine, serotonin, and fatty acids, or brain molecules, used by the gut as signals. It is possible that toxic molecules made by the gut microbiota may lead to the degeneration of enteric nerves, or sensory and sympathetic neurons that surround the digestive system.

An untold number of interactions, ranging from normal functioning to the triggering of inflammatory and autoimmune responses, exist between the vast number of immune cells and the gut microbes.[2] With respect to interaction leading to disease states, we could consider the bacteria an enemy, a force in opposition to the body's state of health. In these cases, when there is a bacterial attack, the immune system is stimulated, and nearby macrophages and other immune cells respond. It is this kind of awareness and action that has led some researchers to call the immune system the "seventh sense."

A hallmark of inflammation is the deployment of immune cells (like soldiers) to the area of damage; upon arrival, macrophage cells engulf and remove bacteria. Cells from the immune system can also surge to sites in the intestine to control bacterial pathogens that invade the gut or prevent widespread activation (an overreaction) of immune cells.[3] With these reactions, the immune system can directly alter bacterial composition in the gut. Some researchers believe that a functioning immune system is necessary to prevent the bacteria from becoming too out of balance in the gastrointestinal tract, a condition called dysbiosis, which can lead to inflammatory bowel syndrome (IBS).[4] Most subjects with IBS also develop Parkinson's disease, supporting a current hypothesis that imbalances in the gastrointestinal system occur years before movement disorders in many disease states.

This leads us to the microbiome, the collection of all of the microbes—bacteria, fungi, and so on—that live in the body, but especially in the gut. Most recently, medical researchers have found evidence (albeit preliminary and mostly from animal models) of a potential role for the microbiome in complex mood disorders, including depression and anxiety, autism spectrum disorder, and schizophrenia, as well as in Parkinson's and Alzheimer's diseases (discussed in later chapters). In other words, an imbalance or a drastic change in bacteria residing in the gut can influence brain development and motor behavior.

Because the gut has a robust ability to sense foreign objects, bacteria often serve as a biomarker, a measurable substance indicative of disease, and a risk factor. Given that there are over a thousand bacterial species, changes in gut bacterial content may be involved in sending signals that affect brain development.

Bacteria can therefore be considered both helpful and harmful in their relationship to the immune system. At times, the immune system can protect gut microbiota, and bacteria in the gut can influence the immune system, maintaining gut health through a system of checks and balances; inflammatory cells secrete molecules that activate the immune system and anti-inflammatory cells that turn down the response. Then, moving beyond mutualism, bacteria and the immune system sometimes have a commensal, or one-sided, relationship that occurs without benefiting or harming the host.

Harmony and Dissonance

The constant give-and-take between the digestive system and its bacterial population, the enteric nerves, and the immune system is like a trio producing harmony. Harmony is the sound of two or more notes played simultaneously. Although determined by many factors, harmony is based on relationships—the joining together—of musical tones. This is also the case in a healthy gut. But when that harmony is not consonant, or when the systems are imbalanced, we have dissonance or a lack of agreement. A breakdown in the periphery or conflict between the systems leads to malfunctions and disease.

Yet there are other theories. One idea postulates that a decrease in molecules like dopamine may also be a trigger for disorders. Dopamine, a neurotransmitter or chemical messenger, influences mood, increasing feelings of reward. It is also responsible for regulating body movements. One can feel the positive effects of this brain chemical by exercising, spending time in the sun, or meditating. On the other end of the spectrum, low levels of dopamine are associated with moodiness and exhaustion, which themselves are harbingers of a disease state.

Although dopamine is a brain chemical, it is also found in the gut, where it plays a role in the development of enteric neurons. And thanks to Frederic Lewy and Heiko Braak's research, we know that Lewy bodies—clumps of protein in the brain that can cause issues with memory, movement, and behavior containing α-synuclein, a small protein expressed primarily in neural tissue and in tumors—can also appear in the enteric nerves. This is another clue that not only is there an important connection between the gut and the peripheral nervous system but also that diseases may actually begin in the gut, or the periphery.

But how could these abnormal aggregations of proteins—the Lewy bodies—and small molecules in the gut spread to the brain? One hypothesis is that they travel through an open passageway that permits travel in both directions, via the vagus nerve.

The Open Sensory Gateway

Recall from the previous chapter that when we were infants, the open sensory gateway between the gut and the brain was critical for developing our bodily systems. A bidirectional communication thoroughfare (Figure 2.1), the vagus nerve runs through the esophagus between the lungs and heart and ends in the brainstem. It is the longest of twelve cranial nerves and is made up of parasympathetic nerves and over two thousand sensory neurons with their cell bodies located at the base of the brain in the brainstem. As an open sensory gateway, the vagus nerve connects the brainstem to the chest and abdomen through branches (exit ramps and roads), merging the brain with the autonomic nervous system and allowing the brain and immune systems to receive signals from the microbiota or bacteria about the state of the body. It is responsible for carrying out involuntary functions, such as controlling food digestion, keeping the heart beating at a constant rate, and providing acidic conditions to sterilize incoming food in the stomach. "Scientists are now focusing on the thinking that happens not in your brain but in your gut," wrote political columnist David Brooks in a recent *New York Times* opinion piece. "The vagus nerve is one of the pathways through which the body and brain talk to each other in an unconscious conversation."[5] By virtue of the thousands of sensory and motor neurons that connect the gut to the brainstem, as well as its role as the main integrator of sensory information, we should consider the vagus nerve a superhighway.

One of the many passengers that move from the gut through a fast lane of information in the vagus nerve to the substantia nigra in the brain is α-synuclein.[6] Although we are still studying the functions of α-synuclein, it provides an example of how a normal protein can

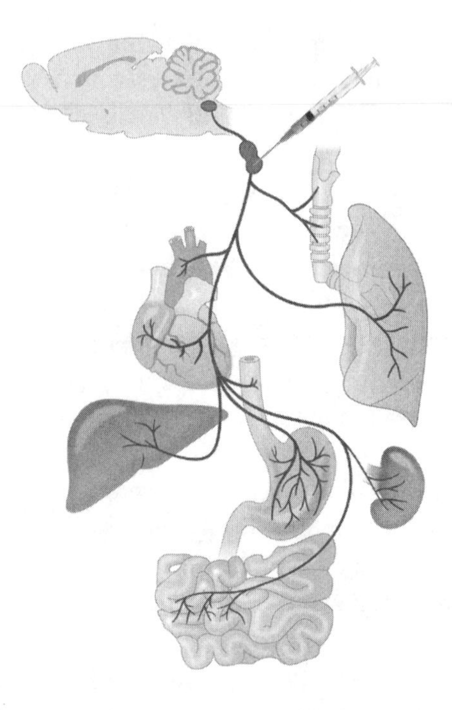

FIGURE 2.1 The paths of the vagus nerve in the periphery. The vagus nerve is considered a cranial nerve, which sends branches to the heart, lungs, esophagus, larynx, and digestive tract, connecting the gastrointestinal tract and the brainstem in the central nervous system.

function when there is balance in the body and what can occur when there is imbalance.

Studies suggest that α-synuclein may help regulate the release of dopamine, creating a pleasurable feeling as part of our reward system. The protein is also associated with an increased rate of neuronal cell death, found in certain neurological diseases. Recent work has also shown that misfolded and aggregated Lewy body proteins first appear in the enteric neurons of the gut, well before they are present in the midbrain.[7] This is another clue, albeit an important one, to support the gut's involvement in neurological disease. Hence, the appearance of α-synuclein in the gut acts as both the messenger and the executioner, an initial indication of things to come and a telltale sign of impending distress to dopaminergic neurons.

Although α-synuclein and Lewy bodies are transported by the vagus superhighway, researchers have found that they also move from cell to cell.[8] Such an idea would have been fanciful a few years ago. Yet researchers have recently observed a release of disease proteins from neurons and uptake by neighboring neurons, supporting the idea that a disease can spread unimpeded from one place to another, with a direct connection to neurological disorders.

We have seen this kind of spread among proteins in other circumstances. Prions, the most well-known example, are considered infectious proteins, as they can be transformed from a normal state to a misfolded contagious state, spreading in an unpredictable manner.[9] In humans, prion proteins have been found to produce rare and (so far) incurable neurodegenerative diseases like kuru and Creutzfeldt-Jakob, often referred to as transmissible spongiform encephalopathy. Brought on by direct exposure and spread to the brain and nervous tissues, Creutzfeldt-Jakob disease occurs in one person in a million, with death within a year of onset of symptoms.[10] Kuru, whose name is derived from the New Guinea term meaning "to shake," was identified in the Fore population of Papua New Guinea and is linked to cannibalism. Both these diseases cause tremors and loss of coordination. Prion disorders found in animals may be a bit more familiar: mad cow disease or bovine spongiform encephalopathy in cows and scrapie in sheep. In the early 1980s, Stanley Prusiner proposed that a single prion protein could not only cause mad cow disease but also multiply and propagate the disease in animals and humans. Trained as a neurologist and a biochemist, Prusiner set a new precedent for how proteins, rather than viruses, RNA, or DNA, cause neurodegeneration, and he revealed how prion proteins can replicate themselves as aggregates of misfolded proteins that lead to cell death.[11] Although not infectious like classic prion proteins in mad cow disease, α-synuclein proteins can spread in a prion-like way, accumulating, misfolding, and spreading from the periphery to the brain.

But there are still connections to be made and mysteries to be solved. Researchers do not know exactly how aggregated proteins spread through exiting and reentering cells. And although some researchers believe that α-synuclein can diffuse throughout the nervous system and can be transmitted from cell to cell, the idea is still theoretical and has yet to be proven. However, it is clear that α-synuclein can cause toxic effects on dopaminergic neurons in distant areas, leading some to believe that α-synuclein can be a clue that things are going awry.[12]

We do know that most sensory messages and other proteins travel through the vagus nerve from the gut, or other parts of the periphery, to the brain. From there, the brain can react. But what if there were even more connections between neuronal cells that further supported the critical role of the periphery, especially in the gut? Recent findings endorse the gut's role as the "brain in the bowel," including that the microbiota (bacteria) in the gut-immune system can directly influence pain-sensing neurons.[13] This is an excellent place to continue our investigation: where our ability to detect pain begins.

3

PAIN PERCEPTION

> We may lay it down that Pleasure is a movement, a movement by which the soul as a whole is consciously brought into its normal state of being; and that Pain is the opposite. We can't learn without pain.
>
> —ARISTOTLE

PERHAPS ARISTOTLE WAS RIGHT: we cannot learn without pain. Yet many of us go to great lengths to avoid it. In fact, the avoidance of pain touches on one of Sigmund Freud's signature ideas, the *Lustprinzip*, or pleasure principle, which proposed that the mind seeks pleasure and aims to avoid torment.[1] While it may be our human nature—and our psychology—to gravitate to pleasure, our bodies evolved an ability to sense and respond to changes in the environment, including injury, without our conscious effort.

It is widely acknowledged that the brain is responsible for the control of pain. After all, the brain is required to respond to harm, as in the classic example of fight or flight. But it has not yet been recognized that it is the periphery that is primarily responsible for pain perception. Although pain may be perceived differently by different individuals, the means by which it is transmitted to the brain is the same for everyone: our sensory receptors, or nerve cells, send messages—which begin in the periphery—to the brain, and then we feel pain. (The brain itself, it should be noted, does not feel pain, likely because the brain does not possess any pain receptors [nociceptors]; it is for this reason, for example, that no pain is felt during craniotomy surgery, despite the fact that the patient is conscious during the procedure.)

Additional studies on pain pathways to the brain have shown how pain is processed and perceived. As discussed, our sensations are controlled by a huge network of nerve endings associated with the peripheral nervous system. Peripheral nerves carry information about pain and temperature through their cell bodies, housed in clusters in the dorsal root ganglia, lying adjacent to the spinal cord and via sensory neurons that extend both to skin areas and directly to the spinal cord. Through these connections, sensory neurons send impulses up the spinal cord to the brain area called the somatosensory cortex. The brain receives and processes pain messages from the periphery; critical communication from the periphery is where pain begins.

After an injury, pain is generated from a complex inflammatory soup, a veritable bouillabaisse of different molecules that are released in the periphery. These immune cytokines, neurotransmitters, prostaglandins, and peptides have the capacity to increase sensitivity to temperature or touch; this collection of proteins and small molecules frequently cause peripheral nerves to be hypersensitive to pain. Breakdown products of metabolism in the liver, such as lactate and neurotransmitters, affect pain transmission.[2] Further, a necessary part of pain detection comes from receptor molecules that lie embedded in the membrane of cells. As discussed earlier, these receptors help us distinguish temperature, pressure and texture, mechanical force, and itch. John Langley, as noted in

Chapter 1, proposed that the enteric system should be considered a third division. He also coined the term *receptive substances* in 1905 to describe receptor proteins in the cell membrane that respond to external chemicals such as growth factors, hormones, and neurotransmitters.[3] This term is an important concept that dominates cell and molecular biology today.

Although pain is usually experienced as an unpleasant sensory and emotional experience, it is protective and provides awareness of potentially harmful events. Therefore, pain can also be considered a form of learning designed to sense changes. It is a detection system that provides an alert to injury and danger. Consider what occurs when our hand touches a hot pan on a stove. We experience the heat—sensation—through a system of nerve endings and touch receptors in our hand that change in response to stimuli. Sensors—thermoreceptors—in the skin inform us about the temperature (hot!), mechanoreceptors communicate pressure and texture (smooth), and nociceptors inform us about pain or injury (ouch!). When our hand touches the pan, the receptors are activated, and the chain of events begins—the periphery feels the pain, and the brain processes the pain signal. The first receptor signals the touch to another neuron, then the next neuron passes the information along to the next, and so on until the message is received by the brain.

To take this a step further, Stephen McMahon, the late Sherrington Professor of Physiology and director of the Pain Consortium at King's College in London, asserted that the periphery is where pain begins *and* that it is central to pain perception. "There is now a large body of evidence," McMahon wrote in 2013, "which strongly suggests that activity from the periphery is essential, not only to initiate but also to maintain, painful symptoms."[4] In this way, McMahon is stating that the peripheral nervous system offers us early warning to pain; like a honk from a passing car, pain is a warning and a safety signal from the periphery.

Types of Pain

To explore this hypothesis further, we must investigate the five most common types of pain: acute, chronic, nociceptive, neuropathic, and radicular. The periphery has a connection with all of them.

Acute pain is severe and short-lived, a signal that there is an injury. Chronic pain is a longer-term experience, a continued communication that there is a problem. A third type of pain is related to the earlier example of touching the hot stove. As noted, nociceptors (from the Latin *nocere* or "to hurt"), one of the thermoreceptors in the skin, detect damage to the body tissue. Another common instance is the throbbing pain that ensues after falling off a bicycle and scraping a knee, which your nociceptors sense and communicate. Pharmaceutical researchers have capitalized on this knowledge to produce pain killers, such as local anesthetics, to block the nerve endings and inhibit their transmission to stop discomfort.

The fourth broad classification, neuropathic pain, arises from damage to the nerves or to a part of the nervous system. Nerves and blood vessels surrounding nerves in the periphery are blocked or damaged, deterioration that can be caused by alcohol abuse, diabetes, or chemotherapy.[5] The resulting misfiring of nerves in the periphery causes neuropathic pain, which has been described as sharp, stinging, and burning.

The fifth type of pain—radicular pain—occurs when the spinal nerve is compressed or inflamed. You may be familiar with sciatica, a pain that issues from the sciatic nerve running from the lower back through the hips and down the leg that is marked by tingling or numbness.

Although a sixth kind of pain—visceral—is not one of the largest classifications, it is noteworthy in having strong ties to the periphery, often originating in the belly. Whereas neuropathic and nociceptive pain come from the neurons or nervous system, visceral pain originates in the internal organs and blood vessels. This type of pain is often felt as a dull ache or a deep pressure that begins in the middle of the body, although it can be experienced other than at the site of the affected area.

Researchers have investigated the connections between malady and an alteration in bacterial composition in IBS, a condition characterized by discomfort and difficulties with regular bowel movements and frequently with pain associated with bowel movements. Researchers have linked the response to hypersensitivity of the nerves that respond to pain (peripheral nociceptive nerves). In addition, the overall feeling of pain suggests that bacteria in the microbiota-gut-brain axis exert a heavy hand (recall our discussion in the previous chapter). Although the identity of causal bacteria is not known, patients diagnosed with IBS often have changes in gut microbiota.[6]

In addition to the five classifications of pain and the less frequently occurring sixth, I believe we also need to consider a seventh: inflammation from tissue injury or infection. I consider inflammatory pain to be an adaptive and protective response. Arthritis, a common condition that often presents as joint pain, stiffness, and swelling, can arise from the loss of cartilage and damage from bones in contact with each other. The body's responses (sensations of discomfort) are protective, alerting us to the potential for further deterioration. In other kinds of inflammatory pain, such as postoperative discomfort, we are responding to noxious stimuli that can occur during an immune response; we feel pain as the immune system activates, mobilizes, and invades, responding to tissue damage. In fact, one definition of inflammation is the mobilization of immune cells to the site of injury. But how do we sense stimuli that are strong enough to threaten our body's integrity?

Sensing Pain

In 1919, scientists made a breakthrough by using capsaicin, the active ingredient in chili peppers. Named after the Latin term *capsa* (meaning "case," in reference to seeds), which itself evolved from the Greek term *kapto* ("to catch"), capsaicin's chemical structure is closely related to vanillin, the primary organic compound found in vanilla beans. For years, capsaicin, taken from the pepper plants' tissues and applied as a cream or gel, had been used as an analgesic to relieve pain. In later studies, researchers found that application of capsaicin to the skin causes burning pain and a heightened sense of pain sensitivity, or hyperalgesia, although this burning feeling and the binding of the receptor to nerve endings does not cause tissue damage.[7]

We owe this knowledge to David Julius, a professor of molecular biology and medicine at the University of California in San Francisco, who, with Ardem Pataputian, received the 2021 Nobel Prize in Physiology or Medicine for the work. Using a clever assay to screen for thousands of genes, Julius was successful in identifying TRPV1, the molecule that binds and responds to capsaicin.[8] A multifunctional protein—a receptor and an ion channel—TRPV1 is expressed by peripheral sensory neurons that detect temperature and pain, and it is sensitive to temperatures that rise above 43°C and to pH levels that decrease below six. In the case of capsaicin, which lowers pH at the site of application below six, TRPV1 is triggered into action and causes the burning feeling.[9] TRPV1 is one of thirty members of a family of transient receptor potential (TRP) ion channels, and although they all sense hot and cold and detect chemicals that are released after tissue injury, TRPV1 is the only one that responds to capsaicin. Because of the ion channel's efficiency in detecting pain from many different kinds of stimuli, it has been an obvious target for pharmaceutical companies to find ways to reduce pain, and researchers have realized that receptors can recognize quite different molecules found in nature, such as menthol, peppermint, wasabi (horseradish), garlic, and even scorpion toxin and cannabinoids, which interact with sensory neurons.

In addition to the molecules that sense temperature and pH, like TRPV1, there are also molecules that perceive mechanical force. Thanks to discoveries by Ardem Patapoutian at the Scripps Institute, we know that another class of ion channels, the piezo receptors (derived from the Greek word for pressure) are responsible for touch sensation and proprioception, or our ability to sense movement, and they have the ability to detect force. Piezo channels form pores in cellular membranes and are found in the nervous system, blood vessels, and the lymphatic system.

Together, the TRPV1 and piezo ion channels serve as the first layer of protection in the periphery, detecting noxious heat and pressure; then our bodies react to the sensation of pain with a defensive response. Nociceptive sensory neurons—the sensory neurons that mediate pain in the periphery—are responsible for recognizing diverse stimuli and setting up a second layer of defense. To see how they do so, let's consider what occurs when we have an itch.

As first described by German physician Samuel Hafenreffer in 1660, an itch (or pruritus) is an unpleasant sensation originating in the skin that provokes the urge to scratch.[10] The sensors in the skin of the forearm are composed of receptor molecules that sense pain, temperature, movement, and itch: nociceptors, thermoreceptors, mechanical force receptors, and pruriceptors. If we have an itch on our forearm, our pruriceptors are stimulated. Then they use molecules to send signals to the brain, from the periphery through the dorsal root ganglia and spinal cord to the higher centers in the brain. The message? Scratch the itch!

Beyond our external reaction (scratching), the messages elicit an internal response as well. The signals result in a release of histamine from mast cells, a type of white blood cell found in the immune sys-

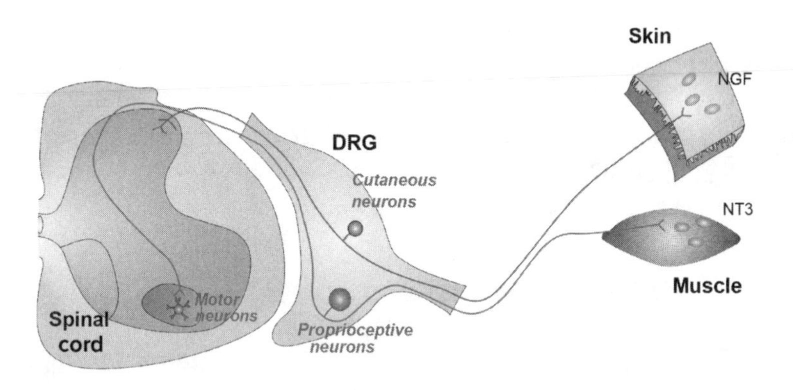

FIGURE 3.1 The DRG (dorsal root ganglion) contains sensory neurons that perceive nociceptive pain and proprioception (balance). The cell bodies of DRG neurons are found in each ganglia up and down the spinal cord. Each cell body extends a process (axon) to peripheral tissues and to the dorsal side of the spinal cord.

tem. (This is the reason we often use antihistamines to counteract itching conditions.) The unpleasant itch can also be suppressed by cold stimuli or through application of menthol through the cold receptor TRPM. Capsaicin is sometimes also used as an anti-pruritic agent to suppress itch, although it requires repeated and prolonged application, since it can activate both pain and itch through its channel, TRPV1.

Figure 3.1 is a visual representation of how the receptors in the skin and muscle use molecules to send signals from the periphery through the dorsal root ganglia and spinal cord to the higher centers in the brain. The dorsal root ganglia is composed of sensory neurons, cell types of different sizes, electrical properties, and gene expression that allow us to feel sensations ranging from pain to pressure and temperature.[11]

The traditional view of the nervous system is that it is rigid and hardwired, since precise connections and circuits need to be made to generate movement and behavior. However, studies of how the peripheral nervous system signals and responds to inflammation, injury, and itch show an inherent flexibility. In particular, studies on pain indicate that there is much plasticity in how signals are generated and modified, including by a protein that is implicated in chronic pain: NGF, the first identified growth factor.

Nerve Growth Factor

Nerve growth factor (NGF) is responsible for the growth, development, and maintenance of sympathetic and embryonic sensory neurons; NGF keeps peripheral neurons alive during development of the nervous system. As dramatic experiments carried out in the 1950s indicated, a lack of NGF results in immediate death of sensory and sympathetic neurons.

The story of NGF's discovery by Rita Levi-Montalcini (1909–2012) is one of persistence and serendipity. Levi-Montalcini, who grew up in Turin, Italy, and went to medical school in the early part of the twentieth century, was one of the few women to be awarded a Nobel Prize (the 1986 Nobel Prize in Physiology or Medicine, which she shared with Stanley Cohen), having endured obstacles due to her gender and her religion (she was Jewish), as well as to politics and war. Although she became interested in basic research in neurology after medical school, with the outbreak of World War II, she was prohibited by virtue of her Jewish heritage from working at any Italian universities. From 1940 to 1943, Levi-Montalcini carried out her research in a makeshift laboratory in her bedroom without any financial support. She used chick embryos to study the effects of target tissue on nerve growth, giving rise to methods that made the detection of NGF possible.[12]

In the 1950s, Levi-Montalcini and Cohen identified NGF, a discovery that led to the burgeoning field of growth factors across neuroscience, immunology, and cancer research. NGF was the first so named as a growth factor, followed by EGF (epidermal growth factor), FGF (fibroblast growth factor), PDGF (platelet-derived growth factor), and many others. Levi-Montalcini concluded that a cancerous cell released nerve growth factor, and she found that sensory nerve cells reacted promptly to the protein, producing nerve fibers. Further, with Victor Hamburger, an eminent embryologist, Dr. Levi-Montalcini made great strides in understanding how the peripheral nervous system is formed. She and Hamburger found that numerous sympathetic and sensory neurons were dependent on the effects of NGF in peripheral ganglia. In other words, the protein, as the name "growth factor" suggests, promoted neural growth in embryos; these proteins were critical in development of cells in the peripheral nervous system.

Because the nature of NGF was still unknown, Stanley Cohen was recruited by Hamburger to determine which of the components were responsible for the activity. The story was remarkable. The expectation, as suggested by the famed biochemist Arthur Kornberg, was that the enzyme—phosphodiesterase from snake venom—would degrade NGF if it was composed of DNA. However, the experiment yielded an unexpected result: snake venom itself contained a large amount of NGF protein, more than the cancerous cell or tumor. Further research showed that not only was snake venom a rich source of the protein, so was the salivatory gland in mice.[13] Through these experiments, they were able to identify the structure of NGF.

Today we know that many cell types synthesize, or produce, hormones or substances that affect their own development, as well as the development of their neighboring cells. This is where NGF comes in—to enhance growth of other cells, an ability that protects and supports survival of peripheral neurons. However, it can also enhance sensations or growth that we consider less than ideal; as Montalcini and Cohen's work showed, NGF can amplify pain and assist the spread of cancerous cells.[14]

Levels of growth factor greatly increase after injury or inflammation. Excessive amounts of the protein can cause hypersensitivity of nerve endings that can produce pronounced pain signals.[15] In the 1990s, clinical trials found that using NGF as a means to reduce the death of neurons during disease or neuropathy resulted in an unexpectedly sharp pain response, thought to be through rapid modulation of heat and pain receptors. But why would a protein that sustains life be involved in producing pain?

Early in the development of the peripheral nervous system, sensory neurons are dependent on NGF to stay alive, and after birth, most peripheral neurons still rely on it for survival. However, in an adult human body, the protein is no longer required for survival; indeed, studies have found that removal of peripheral NGF does not cause death of adult sensory or sympathetic neurons.[16] These observations imply that neurons become independent from growth factors during maturity later in life and that NGF dramatically changes its plasticity during aging.

Over the past decade, there have been attempts to remove NGF from painful areas by using antibodies that block or sequester NGF.[17] These clinical trials have been highly successful, and pharmaceutical companies have been using this approach to combat different kinds of pain. In one study of chronic knee pain from osteoarthritis, physicians treated patients between the ages of forty and seventy-five for between two weeks and six months with an antibody to block NGF. The treatment was highly effective; the patients did not feel any pain.[18] In fact, because there was no sense of pain, many of the subjects overused their joints. Several large pharmaceutical companies, such as Pfizer, Lilly, and Regeneron, are now conducting large clinical trials to test these anti-NGF therapies, targeting different pain conditions using antibodies or small molecules to block one of the molecules in that inflammatory bouillabaisse mix.

In addition to treating osteoarthritis, this approach is also used to alleviate low-back pain and pain produced by cancer drug treatments. Because of the current problems with using opioids and non-steroidal anti-inflammatory drugs as pain medications, the use of anti-NGF antibodies is a potential new approach to the age-old problem of pain management.[19] Reducing NGF shows promise for the treatment of neuropathic and nociceptive pain. NGF increases the development and survival of nerve cells by sending instructions through its tropomyosin receptor kinase A (TrkA). As discussed earlier, the receptor detects pain, inflammation, and many types of injury events by binding near the locale of an injury. In fact, studies have shown that the TrkA receptor can generate hypersensitivity and that pain sensitivity in the periphery is closely related to the NGF-TrkA pathway.[20] On the other side of the spectrum, mice and humans that lack TrkA suffer from dramatic losses of sensory and sympathetic neurons.

And what of patients who lack the ability to sense pain? There may be a clue in the mutations found in the human NGF gene. Researchers discovered that patients with one NGF mutation, referred to as the "painless NGF" gene, have traces of mental retardation and mild cases of anhidrosis, a lack of perspiring.[21] Further, defects or mutations in the TrkA receptor give rise to a dramatic loss in the number of nociceptive fibers, the previously discussed sensory neurons that mediate pain and warn of potentially damaging stimuli.[22] The result is a profound loss of pain sensitivity, damage in limbs, fingers, and toes, or an elevation of body temperature when the body produces or absorbs more heat than it dissipates. As important heat sensors in the skin, TRP channels can sense when the body temperature rises above or falls below 96–99°F (35–37°C) in the daytime. This ability of the body to signal the change in temperature is important in daily living but especially in circumstances like hyperthermia, when one's body temperature can reach 104°F (40°C).

The pain-detection system is a complex sensory experience that varies with tissue damage and with the many associated biological factors, including inflammation, and psychological factors, including attention, emotional state, and expectation. The most common response to injury or illness, triggered by heightened activity in primary sensory neurons, is an obvious, sharp, communication from the periphery that something is wrong.

But sometimes the periphery sends a more subtle signal, a warning of a future disease. With the knowledge of how our body systems communicate and of the signs the body gives when things go awry, we can now combine observations, theories, and findings just as researchers do in practice. In the next chapter, we will investigate a disorder that may be intimated by signals from the periphery years, even decades, before the onset of recognizable symptoms, and we will meet the remarkable physicians who paved the way to its diagnosis.

4

A PRONENESS TO TREMBLING

> The subject of this case was a man rather more than fifty years of age, who had industriously followed the business of a gardener, leading a life of remarkable temperance and sobriety. The commencement of the malady was first manifested by a slight trembling of the left hand and arm, a circumstance which he was disposed to attribute to his having been engaged for several days in a kind of employment requiring considerable exertion of that limb. Although repeatedly questioned, he could recollect no other circumstance which he could consider as having been likely to have occasioned his malady.
>
> —JAMES PARKINSON

IT WAS A MYSTERY. Six otherwise healthy men between the ages of fifty and sixty-five showed similar symptoms: a slow, hesitant gait, severe constipation, difficulty swallowing, and persistent shaking. In 1817, these symptoms, experienced separately, may not have been something to ponder, but since each of the six cases showed all these symptoms, the situation was unusual indeed.

In the passage quoted above from his 1817 "An Essay on the Shaking Palsy," James Parkinson (1755–1824) describes the six cases and, for the first time, documents a nervous disorder characterized by trembling, weakness, and a stooped posture. All these symptoms, and the resulting disorder, involve the periphery. And as we will see, throughout history, misconceptions and mystery surrounded one of these symptoms that was related to the peripheral nervous systems: a proneness to trembling.

London in the early nineteenth century was on its way to becoming a wealthy city, the capital of the British Empire, and the world's largest metropolis, numbering nearly seven million people. Yet nineteenth-century London was also a city of pervasive poverty, with more than a third of its population living in crushing hardship.[1] In this challenging and turbulent time, London's population was ripe for observation. It was in this environment that James Parkinson, then a budding physician, received his medical training, first at the London Hospital Medical College and then at the Royal College of Surgeons.

During his lifetime, Parkinson was known for his contributions as a physician and writer, but it was his keen observational skills that determined his legacy. His ability to study things closely was sharpened by his geological interests as a younger man. He was one of the founding members of the Geological Society of London, responsible for naming several fossils—*Nipa parkinsoni* and *Nautilus parkinsoni*—and he had the distinct honor of having an ammonite fossil named after him, *Parkinsonia dorsetentsis*.[2] Parkinson's close study of fossil records and the evolution of organisms over time translated easily to medicine.[3]

In the London neighborhood of Hoxton, Parkinson spent time observing his patients and watching people walk down the street. The man whom Parkinson described in the passage quoted above—a gardener—stood out from other passersby by his short, shuffling steps, his balance problems, and his interminable shaking. Parkinson became fascinated by the halting movements, which he began to observe in patients and other pedestrians. Over time, Parkinson noticed that these types of movements often developed into what he called the "shaking palsy," a disorder characterized by muscular rigidity and weakness and tremor.[4]

The study of neurological disorders was in its infancy at this time; there were no accounts of motor neuron diseases or other diseases that had these symptoms. But through his extensive clinical experience and discriminating observational skills, Parkinson knew that the shaking palsy was a medical affliction, not just a consequence of normal aging. For this reason, he recorded his observations.

From Galen to Poussin

Although Parkinson is known for his observations, he was not the first to witness this kind of "involuntary tremulous motion." There were many noted examples of uncontrolled shaking throughout history, many dating as far back as the Old Testament. Perhaps the first on record, Galen (129–99), a famous Greek physician during Roman times, noted the differences between the sensory and motor systems and made many references to tremor, gait disorders, and paralysis, which he believed developed from mental impairment and muscular instability.[5] But shaking was also referenced in popular culture as a subject of interest for many artists and writers, including some of the greatest and most well known today: Leonardo da Vinci, William Shakespeare, and Nicolas Poussin.

Leonardo da Vinci (1452–1519), famous for his studies of motion and action that featured an accuracy worthy of an engineer, was an artist and a scientist. As Walter Isaacson wrote in his fascinating biography *Leonardo da Vinci*, da Vinci was notably interested in how the brain and the nervous system translated into movements of the body.[6] This enthusiasm is apparent in his documentation of and insight into involuntary movements: "Nerves sometimes operate by themselves without any command from other functioning parts or the soul," da Vinci wrote. "This is clearly apparent for you will see paralytics and those who are shivering and benumbed by cold move their trembling parts, such as their heads or hands without permission of the soul. . . . Its forces cannot prevent these parts from trembling."[7]

The brilliant playwright and poet William Shakespeare (1564–1616) was also a brilliant observer of reactions of the human body, from bipolar behavior to other conditions and symptoms, including problems with sleep and tremor, detailing diseases of the nervous system two centuries before James Parkinson made his observations.[8] Whether Shakespeare was describing vertigo, a change in gait, or an irregular heartbeat, many of his written works, especially the plays *Henry IV, Troilus and Cressida, Richard II, The Tempest, Julius Caesar, Macbeth*, and *A Winter's Tale*, include observations that bear remarkable resemblances to neurological disorders that were later studied by the groundbreaking nineteenth-century neurologist Jean-Martin Charcot.[9] In *Henry VI,* for example, a rebel named Dick asks Lord Saye, a treasurer in the government who advises King Henry of England, why he has a tremor:

> *Dick:* Why dost thou quiver, man?
> *Lord Say:* It is the palsy, and not fear, provokes me.[10]

References to shaking palsy can also be found in *Troilus and Cressida*, a play about the Trojan War. The Greek commander Ulysses described the condition of Achilles, his lionized fellow Greek warrior, who was suffering the ravages of time—and also a shaking disorder:

> And the, forsooth, the faint defects of
> age Must be the science of mirth; to
> cough and spit
> And with a palsy fumbling on his gorget
> [armor] Shake in and out the rivet.[11]

Similarly, in *Richard II*, written in 1595, the aging Duke of York is asked to take the place of Richard II, with another reference to palsy:

York: How quickly should this arm of mine
Now prisoner to the palsy, chastise thee.[12]

Shakespeare also frequently used syncope—fainting or passing out, a brief loss of consciousness—as a dramatic device in his plays. Syncope can be caused by a heart problem, by a psychiatric or neurological condition, by standing up too quickly along with decreases in blood pressure, or, in Shakespeare's plays, by the sight of blood. Today we know that this is another symptom of a breakdown of the peripheral nervous system.

Nicolas Poussin (1594–1665), a prolific French Baroque painter, experienced shaking palsy firsthand. Before his movements slowed in his later life, Poussin produced a vast output of landscapes and mythological and religious paintings that influenced many artists, including Jean Auguste Ingres, Paul Cézanne, Georges Seurat, and Pablo Picasso. Poussin suffered from a regular persistent tremor, which made drawing and painting difficult. An in-depth analysis of his paintings was later conducted by researchers at University College in London. Two professors, Patrick Haggard and Sam Rodgers, concluded that Poussin experienced a tremor early in his life, from age twenty-six, that worsened with time.[13] By examining the tendency to freeze before each stroke, they were able to follow each stroke in the drawing and measure the movement velocity by observing how the ink accumulated on the painting. They also took into account the number of ink blots that collected on the canvasses, likely a result of tremor. A plot of the magnitude (amplitude) of the tremors from over fifteen images showed that tremor occurred more frequently with increasing age (Figure 4.1).

Shaking and other symptoms were clearly on display well before James Parkinson's first medical observations. But it was his careful, detailed study of the tremors, slow body movements, hunched posture, sleep disturbances, and difficulty swallowing, as well as his awareness

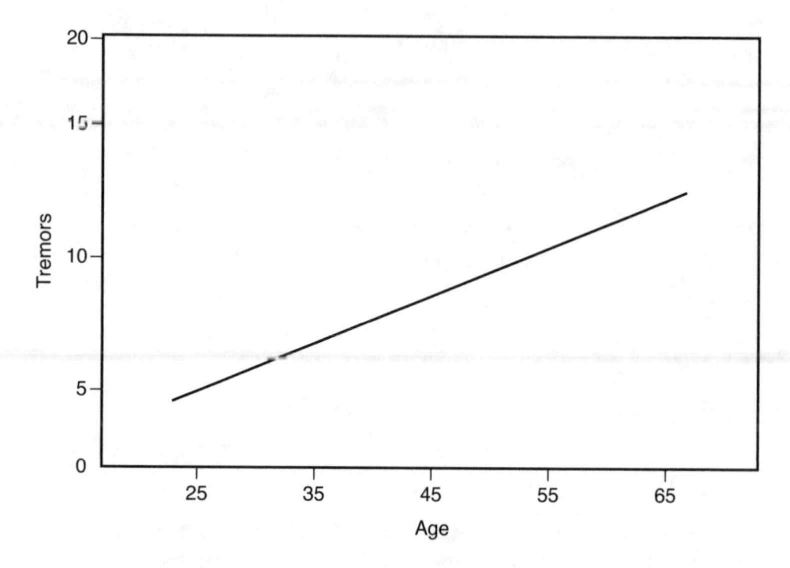

FIGURE 4.1 An increase in Poussin's tremors was detected with age. The strength of Poussin's tremors was measured in his paintings as a function of age; x-axis is the date of Poussin's painting; y-axis is the strength of tremor, calculated from painting strokes of fifteen drawings.

of the gradual escalation of the symptoms, that allowed Parkinson to fully understand both the symptoms and the disease.

The Observations

Parkinson found that the patients experiencing these symptoms tended to be older and that many early symptoms of the disease became apparent at least ten years before the onset of visible tremor. Noticing the characteristics of the disease at an early stage, in addition to focusing on symptoms later in life, offered Parkinson clues of how the disease progressed:

> So slight and nearly imperceptible are the first inroads of this malady, and so extremely slow its progress, that it rarely happens, that the patient can form any recollection of the precise period of its commencement. The first symptoms perceived are, a slight sense of weakness, with a proneness to trembling . . . most commonly in one of the hands and arms. These symptoms gradually increase in the part first affected; and at an uncertain period, but seldom in less than twelve months or more, the morbid influence is felt in some other part.[14]

Parkinson found that pain was an early symptom; patients experienced not only stiffness but also muscle contractions and balance problems. Then the signs of tremor began, gradually spreading to other limbs and eventually leading to a progression of other movement problems, including a slow gait. Parkinson also asked the patients under his care questions that were not typical of medical exams at the time, including questions about their experiences eating or having bowel movements. He also asked patients whether they had a disruption of sleep, for Parkinson began to realize that issues with insomnia often accompanied the slow gait and trembling. As Parkinson predicted, the patients noted difficulty eating, moving their bowels, and sleeping,

In time, obvious signs of the disorder, besides the tremor, appeared. One symptom was a stooped posture with the trunk bent forward, which was often accompanied by a festinating (hurrying) gait, perhaps a response—subconscious or conscious—to compensate for the body's growing tendency to stoop and shuffle. There was also loss of facial expression—a blank expression or a mask-like appearance in which the facial muscles became locked—and a slowness to smile. Lastly, patients eventually had difficulty with fine motor skills, such as handwriting. As described by Wilhelm von Humboldt (1767–1835), a German diplomat, renowned humanities scholar, and the founder of Humboldt University of Berlin, who suffered from this symptom:

> There occurs trembling or a situation I prefer calling clumsiness rather than weakness. Writing, if it is to be firm and clear requires a lot of sometimes very minute and hardly noticeable finger movements that need to be made in rapid sequence but with clear distinction from each other. In aging, suppleness is missing in this respect. The same applies also to other acts such as buttoning up during dressing, etc, while the hand maintains its strength for grabbing, carrying, holding.[15]

It was a feat for Parkinson to observe, note, and follow these symptoms in his patients over time. But even though there were acknowledgments of Parkinson's efforts during his practice from 1817 to 1861, the medical community did not give him credit until much later in the nineteenth century. It was up to the French neurologist Jean-Martin Charcot (1825–93), born a year after Parkinson's death, to realize that the disorder Parkinson had observed was a specific, but also prevalent, medical condition.

A Distinctive Disorder

Jean-Martin Charcot was the most famous French physician of his time, and the scope of his accomplishments is matched by few neurologists to this day.[16] He is widely credited with establishing the discipline of neurology and with realizing that the shaking palsy described by Parkinson decades earlier was a distinctive disorder.[17] As professor of diseases of the nervous system at the Medical Faculty in Paris, he also diagnosed and named at least twelve neurological disorders, including Charcot-Marie-Tooth atrophy, Tourette's syndrome, amyotrophic lateral sclerosis (ALS), and multiple sclerosis, all diseases that we now know involve the periphery.

Charcot's success as a clinician and a neuroscientist, like Parkinson's, can be attributed to curiosity and the unique skills honed by a hobby; in Charcot's case, drawing. Charcot had the ability to observe, then draw, his patients' posture and physical changes with such precision that he was able to make connections between different diseases. Charcot acquired the ability to remember details from his pictures of the anatomy and pathology of the nervous system. For example, Charcot noticed that many of his subjects suffered from deterioration in movement and muscle coordination. After he observed and drew many of these patients over time, he saw the degradation in their range of motion. He described the condition using the term *sclerosis*, derived from the Greek word for hardening, which could be linked to scarring or muscle weakness.

As is likely clear from the naming conventions, Charcot was the first to diagnose multiple sclerosis as a specific condition in a living patient. As an episodic disease of the central nervous system, multiple sclerosis results in the destruction of nerve fibers' covering (myelin), creating deficits in both nerve and muscle in the periphery that affect vision, balance, and mobility. Although today the condition is detected by high-resolution magnetic resonance imaging, Charcot relied on his neuroanatomical expertise to diagnose the disorder through studying plaques, a residue of cellular tissues that appear as dark gray areas in the spinal cord and brain. He also traced the location of ALS to the spinal cord and detected problems in the motor neurons that emanated from the spinal cord. Because he was able to piece together the features of progressive muscle weakness and the involvement of motor neurons, Charcot is considered the father of ALS disease.[18]

But the symptom of tremor tickled Charcot's curiosity, especially when he realized that some of his patients with multiple sclerosis shared the symptoms of muscle weakness and tremor.[19] Through detailed measurements of the frequency and intensity of his patients' tremors, he found that the shaking in patients with the diagnoses of Parkinson's disease and multiple sclerosis to be relatively slow—four to six oscillations per second. He also found that patients with multiple sclerosis did not experience tremor during rest—it occurred only after some physical activity—and that the oscillations tended to increase with time. On the other hand, patients with Parkinson's disease displayed a noticeable tremor both at rest and during activity, which did not change with more activity. These observations allowed Charcot to differentiate tremors found in Parkinson's and those found in multiple sclerosis diseases.[20] Through these kinds of studies and measurements, Charcot drew several conclusions: abnormal movements were a failure that arises from tremors; in the early phases of the disease, extra movements could actually counteract and suspend the tremor, and in later stages of the disease, the tremor became continuous. Further, he deduced that shaking palsy tremors were related to problems not only with movement but also with posture. He was able to make a distinction between the patients who exhibited severe tremors but no rigidity and the patients who displayed a rigid stance without tremors, and he described the tremor of the hands to be oscillating "in almost a pathognomonic manner. The fingers approach the thumb as if to spin wool, and simultaneously the wrist and forearm flex to and fro."[21] By examining a large number of subjects, Charcot helped establish a basis for diagnosing Parkinson's disease.

As an admirer of Parkinson's 1817 essay, Charcot knew that although Parkinson's study was monumental, it was based on only a few individuals. The next step was to put together a larger study. At the Pitié-Salpêtrière Hospital in Paris, Charcot found many elderly subjects who resembled Parkinson's description, so he devised a simple test to see if head nodding could be distinguished from the shaking characteristic by using a headband with a feather. After carrying out that experiment and other measurements of tremor and signs of rigidity, Charcot grouped his patients into different categories and separate neurological symptoms from just a tremor. In 1887, in honor of Parkinson's original observations of shaking palsy, Charcot graciously christened the disorder "Parkinson's disease," more than six decades after Parkinson's death in London.

The Clues

Today we know that the periphery sends messages about sensory changes or disturbances that can result in symptoms of Parkinson's disease, including tremors, slowness of movement, and sleep disturbances. We know that motor nerves send information from the central nervous system (the brain and the spinal cord) to the muscles and glands throughout the body. And although we are just beginning to grasp exactly how the periphery is responsible for all the vital connections and wiring outside the central nervous system and how it affects blood flow, heart rate, oxygen exchange in the lungs, and even the constriction of the eye muscles, it is clear that the periphery plays an important role in communication.

With powerful new techniques such as RNA sequencing of single cells—methods that rely on the ability to determine the gene expression profile and properties of single cells that reside in the periphery—we have recently been able to identify individual cell types and to uncover differences with neighboring cells.[22] Researchers found that the neuronal diversity, and thereby the functions, are spatially different from those of cells found in the brain, an important indication of regional and functional specialization (as stated in the previous chapters). Suffice it to say, this differentiation explains the periphery's and the brain's different functions, and when combined with Dohrn's evolutionary theory (that the periphery evolved before the brain), we can see how the periphery has an unquestionably critical role.

The Gut Connection

Two of the hallmarks of Parkinson's disease are the loss of dopaminergic neurons in the basal ganglia (located within the cerebral hemispheres of the brain) and the enrichment of the α-synuclein-protein-containing Lewy bodies (clumps of protein that form in the brain). Researchers have found that patients with the disease also had α-synuclein in the periphery or the enteric nervous system.

A current theory, originally postulated by Hiko Braak, asserts that Parkinson's disease begins in the gut (the periphery) and then transfers to the brain, eventually leading to degeneration.[23] Recent research from Braak's group supports this, as patients with Parkinson's had Lewy bodies in the enteric neurons of the gut.[24] Given what we know about the body's ability to communicate and transfer materials from the gut to the brain via the vagus nerve (Chapter 3), it would make sense that the proteins began in one area of the body and transferred to another. We have also specifically seen that the protein α-synuclein travels through the vagus nerve to the basal ganglia and causes the loss of dopaminergic neurons. Further, a fascinating study in Denmark showed that a group of over five thousand subjects who underwent a procedure to remove the vagus nerve—a vagotomy—had a lower risk of Parkinson's disease than the same cohort in the general population.[25] In other words, shutting down the passage from the gut to the brain decreased the incidence of the disease. Researchers also learned that even after removal of the vagus nerve, the enteric nervous system continued to operate—a brain in the bowel, indeed! All of these findings suggest that the transport of a protein from the periphery may result in the spread of Parkinson's disorder.[26]

Along these lines, we must also consider several dramatic studies on the appendix and its removal. It is generally well known that the appendix, which sometimes becomes inflamed, causing pain and requiring surgical removal, is a part of the digestive tract that attaches to the colon. Less well known is the fact that the human appendix is a vestigial organ that contains detectable amounts of α-synuclein aggregates, many of whose forms, we now know, are precursors to those that become pathogenic, or cancerous. A Swedish registry that followed an astounding 1.6 million people for nearly fifty years came to a startling conclusion: removal of the appendix resulted in a lower risk of Parkinson's disease.[27] Further, not only was the individual's risk reduced, but for those who received an appendectomy, the age of onset for Parkinson's was delayed.[28]

Now, this surprising result may be a by-product, a secondary effect or epiphenomenon, and may have nothing to do with the disease. On the other hand, there is a clear correlation, and data from additional studies suggest that this is not an irrelevant finding. First, in individuals both with and without Parkinson's disease, α-synuclein aggregates accumulate in the appendix before the onset of the disease.[29] Second, the appendix is brimming with bacteria, an observation that correlates to the problems in the gut that precede other symptoms of Parkinson's disease, including tremors, by an average of almost ten to twenty years. But to attribute these effects to bacteria, it is necessary to determine if changes in microorganisms are a cause or an effect.

Recall from Chapter 3 that we have trillions of bacteria in our bodies, especially in our gut. In the case of Parkinson's disease, bacteria serve as a biomarker, a measurable substance indicative of the disorder, and a risk factor. The constituents of fecal and gut microbes change from patient to patient, as one would expect. However, researchers have found that mice predisposed to develop symptoms like those of Parkinson's disease have high levels of α-synuclein. An experiment that reduced bacteria in the gut by using germ-free conditions in mice diminished their motor abilities (grip, climbing, walking on a beam), suggesting that it is the range of bacteria in our guts that is a key to gut bacteria's beneficial systemic effects.

Let that idea sink in for a moment: one's movements may be affected by the type of bacteria that grows in the gut.[30] Although these observations are only correlations at this stage, researchers have also found that changes in gut bacterial content may be involved in sending signals that affect brain development.[31] So Parkinson's disease, still considered a neurological disorder (brain disease), has some very interesting connections to the gut (the periphery).

The Pain Connection

Early in his investigations, James Parkinson noticed that pain was one of the first signs of a problem in the subjects he followed.[32] More recent studies support this observation. In patients with Parkinson's disease, painful sensations often occur in peripheral locations such as muscle and bone, and these signals began in the periphery before the onset of other symptoms like tremor. In other words, the periphery is where it all begins.

All this anatomical information and theoretical discussion is well and good, but what does this look like in a disorder, specifically, in this case, in Parkinson's disease? Let's take a look at two cases of well-known individuals to see how the disorder progresses and, perhaps, uncover clues that may have been overlooked. In their stories of diagnosis and eventual decline, we might just find crucial evidence.

Chairman Mao Zedong

Mao Zedong (1893–1976) served as the chairman of the Communist Party in the People's Republic of China and launched the Cultural Revolution. He oversaw a campaign of repression called the Great Leap Forward, which was an effort to reconstruct the country into a communist society.[33] Unfortunately, the effort was marked by catastrophe. Between 1958 and 1962, tens of millions of people died of starvation or torture. It was the largest non-wartime mass death (or killing) in history. Mao aimed to leave a powerful legacy, showing strength in every area possible, through mental coercion and physical exertion. His efforts were successful, as few other authoritarian leaders possessed as much power over so many people. Yet just a few years after one of his famous physical demonstrations of vitality—swimming across the swift and strong currents of the Yangtze River—Mao was never again seen standing or walking in photographs. He formally stopped seeing visitors in 1976, except for ex-president Richard Nixon, who paid him a private visit in China.

In the book *The Private Life of Chairman Mao*, Dr. Li Zhisui, Chairman Mao's personal physician from 1954 to 1976, described Mao's political, personal, and medical history and his eventual decline. Despite his frequent examinations by neurologists and specialists, Mao was suspicious of the medical world and did not accept most of his doctors' opinions.[34] As a smoker, he had some typical damage to his lungs, and over time, his eyesight became weak, and he was diagnosed with cataracts (to accommodate Mao's failing eyesight, special presses were commissioned to print characters with large fonts). Then, in the last two years of his life, Mao became blind, unable to read or to sign documents.

Since Mao had multiple health issues, his case was far from simple. Yet we can see some familiar signs that there were issues connected to his peripheral nervous system. For example, Mao began to show signs of slowness of movement and postural instability, frequently sitting with his head set back against the top of the chair. The muscles in his arms and legs atrophied, and his speech became noticeably distorted and incomprehensible. Neurologists who examined Mao suspected he suffered from Parkinson's disease and ALS.[35] When his motor neurons underwent a slow and progressive degeneration, first resulting in paralysis on his right side, which immobilized him, then extending to his throat, pharynx, and tongue, ALS was confirmed.

Mao eventually became bedridden and suffered from bedsores. The doctors who examined Mao in his last two years knew that his condition was incurable. Because he was afflicted with so many problems, the cause of his death at eighty-two was ambiguous, although the muscle problems reflect a failure of motor and sensory neurons.

But as noted, events that precede the disease are frequently overlooked. In Mao's case, there were early signs of health problems—clues from the periphery—years before his health visibly declined. As a young man in his early thirties, Mao suffered from constipation, moving his bowels only once a week. He also had insomnia. These symptoms are precursors to Parkinson's disease—or in James Parkinson's accounts, "shaking palsy"—and are clearly connected to the peripheral nervous system.[36]

As mentioned previously, the periphery regulates digestion through a push-and-pull system of autonomic nerves. Recall that the autonomic system—the first, second, and third violins—switches from being active to being inactive, in many cases simultaneously, according to needs in the body. Responsible for sensing the state of the organs and the external environment, the autonomic system ensures that our circulatory, pulmonary, reproductive, and digestive systems are in sync. When there are difficulties like dizziness, fainting, constipation, and even insomnia, these incidents in the peripheral nervous system show us that there may be greater issues to come.

Today, physicians who treat Parkinson's disease indicate that dizziness, fainting, and pain often occur at least ten years before the appearance of tremors, or other more recognizable symptoms of the disorder. The same timeline applies to constipation issues, a problem caused by a degeneration or dysfunction of nerve fibers regulating the contraction of muscles in the colon. We also see a similar early warning with sleep issues, especially insomnia. In fact, sleep disturbances are very common in many neurodegenerative diseases, including multiple system atrophy, a combination of diseases that share similar symptoms with Parkinson's.[37] Disturbances in sleep can, in turn, produce peripheral problems, such as pain, depression, and dementia. Although the exact causes of sleep disorders are unknown, a lack of rapid eye movement (REM) sleep has been linked to problems with physical activity and movement, as well as appetite and eating disorders, all symptoms and issues with a peripheral connection.[38]

Researchers believe that sleep is regulated in the brain due to the rhythms of electrical activity, but recent research has revealed a number of genes that regulate the sleep/wake pattern. One of the so-called clock genes is Bmal1 (Brain and muscle factor 1), a gene that drives rhythmic cycles and circadian control; the Bmal1 protein controls the circadian clock. For this reason, researchers expected that Bmal1 would be found primarily in the brain. However, they found that Bmal1 is mostly *outside* the brain.[39]

The Bmal1 protein is expressed peripherally in skeletal muscle, where scientists believe it is able to determine the total amount of sleep far better than if it were located in the brain.[40] How in the world does skeletal muscle influence sleep? We do not yet know, but it is possible that factors like Bmal1 or other related proteins in the periphery alter sleep patterns. Perhaps events outside the brain can control circadian rhythms and sleep patterns, or at least serve as an early warning about sleep issues from the periphery.

If Mao's physicians had the knowledge of our collective research and an awareness of the signals from the periphery, they may have been able to diagnose and treat his ALS and, perhaps, Parkinson's disease. After all, there were plenty of early clues.

Muhammad Ali

Muhammad Ali, like Mao, was plagued by numerous health issues and showed several now-familiar symptoms. Ali's diagnosis of Parkinson's disease, however, came when he was only forty-two. From his start as a boxer in 1960, when he won a gold medal at the Olympic Games in Rome, until 1981, when he retired, Muhammad Ali won fifty-six of his sixty-one professional fights and was crowned heavyweight champion three times. It is not a stretch to claim that his fighting career globalized the sport of boxing, from the United States and parts of Europe to other areas around the world, including Zaire, the Philippines, Malaysia, Iran, Saudi Arabia, and Japan. Ali was famous as a supreme athlete with a masterful blend of power, velocity, and courage.[41]

Outside the ring, Ali also possessed the courage to live his life on his own terms and to be outspoken about his views, even when they were controversial, as when in 1964 he converted openly to the Muslim faith. He became a conscientious objector, and despite the government's rejection of his application for conscientious objector status, he refused to serve in the Vietnam War. Convicted of draft evasion, Ali was sentenced to a five-year prison term and banned from boxing for three years, from 1967 to 1970, during the peak of his athletic career. Although his conviction was later overturned by the US Supreme Court in 1971, it left Ali with a career that was both galvanizing and highly stressful.[42] After his exile from fighting, Ali continued a heavy schedule of bouts against formidable opponents, such as Sonny Liston, Floyd Patterson, and Joe Frazier. Although Ali won hard-fought, brutal matches, his bouts, with Frazier in particular, took a heavy toll on him.

The constant strain in Ali's life and career as a boxer affected his body and his health. His chronic stress elevated the stress hormones—glucocorticoids—that circulated throughout his body. We now know that long-term release of high levels of cortisol has profound negative effects on digestion, anxiety, sleep, and cognition. These precipitating factors likely signaled, or perhaps even brought on, early stages of Parkinson's disease.

The basketball star Kareem Abdul-Jabbar was a staunch supporter of Ali's career. As a present-day activist, Abdul-Jabbar made the observation about sports and stress in life that "whether it's basketball or boxing, there's something . . . about being in a confined space and time with people trying to impose their will on you while you not only fend them off, but impose your will on them. It's pretty much the essence of life."[43]

Then there was the physical toll of boxing itself. Ali's last two fights against two competitive fighters, Larry Holmes in 1980 and Trevor Berbick in 1981, were attempted comebacks that failed miserably.[44] Against Holmes, the fight ended in the eleventh round after a brutal beating, during which Ali suffered severe blows to his abdomen. It was the only fight that Ali lost by a knockout. This beating by Holmes, along with all the other fights, has also been proposed as one of the events that may have led to Ali's eventual decline. The last fight, in December 1981 against Trevor Berbick, was difficult to watch, as Ali lost in a tenth-round decision. A short time after his last fight, Ali was diagnosed with Parkinson's disease.

Three years after his retirement, in 1984, Ali showed many classic signs of the disease: his posture was stooped, his hands trembled, his speech slowed, and he often slurred his words. Instead of bouncing on the balls of his feet and moving quickly in the ring, he was shuffling slowly, flatfooted. The symptoms, mild at first, quickly worsened. He lost his coordination and experienced drowsiness and uncontrolled tremors. He looked frail during his public appearances. Before too long, his neurological condition put Ali on the ropes. Although there was no debate over Ali's diagnosis as there had been in Mao's case, there is continued discussion over the causative factors that precipitated Ali's decline and the onset of Parkinson's. Speculation has focused on four: blunt trauma, environmental factors (specifically toxic chemicals), endogenous substances (chemicals produced within Ali's own body), and genetic mutations. Let's call them the four punches precipitating the knockout.

Some believe that the blunt-trauma injuries from boxing precipitated Ali's decline, largely supported by the growing body of evidence linking traumatic brain injury and neurological decline. Others believe that Ali may have suffered from the cumulative effects of chronic traumatic encephalopathy (CTE), a condition that can lead to dementia. First described in 1928 as "punch drunk syndrome," CTE can be diagnosed only after death, as autopsy samples of brain tissue are needed to make a definitive judgment. A microscopic examination of the autopsied brains of 111 American football players found clear signs of CTE. (American football is another sport where athletes frequently suffer injuries, especially to the head.) Yet there are other symptoms of CTE that can hint at the syndrome, including declining memory, confusion, fits of anger and emotional outbursts, aggressive behavior, and depression.

A prominent research group at Boston University, headed by Ann McKee, Robert Cantu, and Robert Stern, has pursued the diagnosis of former football players between the ages of forty-five and seventy-four. Their studies contributed many insights about the effects of traumatic brain injury from car accidents, military action, and football that accompanied memory loss, personality changes, depression, and problems in impulse control.[45] The CTE Center in Boston indeed found protein deposits and visible lesions in postmortem cases of CTE similar to what has been seen in cases of Alzheimer's disease. A progressive degeneration of brain tissue with an abnormal aggregation of disease proteins, including the protein tau—designated by the Greek letter τ—were commonly seen. Tau forms the internal cell structures of neurons and aggregates in clusters that deposit during aging and neurodegenerative disorders.[46]

Proteins like tau can become dangerous when they change their conformation and become packaged up in bundles ("tangles"). Researchers have found that in the brain, tau-containing tangles are involved in many types of dementias and other neurological diseases. In this altered form, tau disrupts communication between nerves, leading to problems in memory and judgment and to aggressive behavior, anxiety, and depression. Researchers now consider dysregulated tau a signature of many neurodegenerative diseases, including Parkinson's.[47] While there is a strong connection to football injury, a causal connection between boxing and dementia pugilistica has not been entirely established, despite strong circumstantial evidence. But could there be a causal link between trauma to the head and neurodegenerative diseases like Parkinson's and dementia?

A large study was taken of seven thousand elderly individuals who had suffered from a head injury accompanied by a loss of consciousness.[48] Another, separate group consisted of a cohort of Catholic nuns and priests in Chicago, aged sixty-five and older, who were followed for decades. This clinical research, called the Religious Orders Study, was initiated by the Rush University Medical Center.[49] All the participants underwent careful cognitive examination during their lifetimes and volunteered to donate their brains at death. As with Charcot's postmortem deductions, researchers were able to learn about the effects of head trauma from the bodies, along with detailed mental records of each individual over time. Results from this study and several others showed a greater risk of Parkinson's disease after a prior head injury, even more so than for Alzheimer's disease.[50] In fact, head injury trauma was shown to increase the risk of developing Parkinson's disease later in life by some 300 percent.[51]

But these head-injury statistics tell only part of the story of Mohammad Ali's decline, and it makes sense to consider other factors and potential causes. Thus we turn to the second "punch" delivered to Ali's system: the effects of the body's battle with toxic chemicals and its connection to the periphery. Pesticides, insecticides, and herbicides have been blamed for Parkinson's disease for some time; the herbicide paraquat, for example, an effective killer of weeds, has been shown also to be an effective killer of dopaminergic neurons.[52] It has been conjectured that Ali may have been exposed to toxic chemicals, herbicides, and pesticides—all of which can cause misfolding and aggregation of proteins—during his years of training in the Pennsylvania countryside.

Rotenone, another environmental toxin and pesticide, causes neural damage. Unlike the aforementioned chemicals, rotenone does its damage by inhibiting respiration carried out by the mitochondria, which is the energy source in all cells. Rotenone also promotes α-synuclein in the peripheral nervous system, specifically in enteric neurons. These observations suggest that the damage from this chemical may affect other organs and systems, in addition to vital brain areas.

Another chemical of interest is MPTP, short for the complicated chemical name of 1-methyl-4-phenyl-1,2,3,6-tetrahydropyridine, a compound that was found as a contaminant of heroin in the 1980s. For the purposes of this book, the chemical is closely related to paraquat (the weed killer) that produces death in dopaminergic neurons.[53] MPTP rapidly crosses the blood-brain barrier to reach the periphery, destroying dopamine neurons in the enteric nervous system and compromising the gut's ability to carry out gastrointestinal motility. Experiments using the chemical MPTP showed a loss of 40 percent of dopaminergic neurons in the enteric system in mouse models of Parkinson's disease, accompanied by changes in motility in the colon.[54] Further, researchers found that the use of MPTP as a drug created a profound neurological condition very similar to Parkinson's disease, including the symptoms of body rigidity and tremor. But despite knowing that MPTP kills dopamine-producing neurons, we still do not know the exact originating causes.

The third theoretical punch in Ali's bout with potential causative factors for Parkinson's may have been delivered by chemicals residing within his own body. We have previously encountered dopamine as the "feel good" hormone that is part of the hardwired system that causes us to seek out rewarding behaviors that release dopamine into our bloodstream. Dopamine is also the key neurotransmitter—a chemical messenger—that allows communication between nerve cells to control movement. While today it has been established that a lack of dopamine is associated with movement disorders, Arvid Carlsson (1923–2018), an acclaimed Swedish pharmacologist, proposed that dopamine, the critical brain chemical that passes signals between neurons in the brain, is responsible for sending messages in the brain and the periphery.[55] He also asserted that dopamine could be responsible for Parkinson's disease.

To test his theories, Carlsson performed an unconventional experiment. He decided to use a potent drug called reserpine, which is derived from plants and is frequently prescribed for blood-pressure control and as an antipsychotic, since it lowers dopamine levels. Carlsson's experiment involved dosing rabbits with reserpine, which caused a severe loss of movement. Then, to restore the rabbits' mobility, he administered L-dopa, a drug that is converted into dopamine in the brain. During the course of the experiment, Carlsson realized that the movement difficulties of his rabbits were similar to those of people with Parkinson's disease. He then proposed that the illness was related to a loss of dopamine. Following Carlsson, other scientists confirmed that dopamine is depleted in people with Parkinson's disease, and L-dopa soon became the standard therapeutic treatment for the illness.

Carlsson extended his observations to other drugs, including cocaine, which strengthens dopamine signaling, and he contributed to research on serotonin, another neurotransmitter, which led to the development of widely used antidepressants such as Prozac. For this research, Carlsson was awarded the Nobel Prize in Physiology or Medicine in 2000, which he shared with Paul Greengard and Eric Kandel, who both made fundamental advances in how communication occurs between neurons.

Since Carlsson's time, researchers have found that in Parkinson's disease, dopaminergic neurons are abundant in one of the structures called the substantia nigra, a small region residing in a group of structures embedded deep in the brain—called the basal ganglia—whose job is to produce dopamine. As part of the basal ganglia, the substantia nigra plays an important role in regulating muscle control, balance, and movement. The dopaminergic neurons in the substantia nigra produce and release dopamine to communicate with other segments in the basal ganglia.

Let's look at this incredible communication process a little more closely. Neurotransmitters—the chemical messengers—are stored in small spherical packets within nerve cells called synaptic vesicles. A nerve impulse at nerve connections (synapses) causes an influx of calcium in the presynaptic membrane, containing neurotransmitters in the vesicles, located at the end of an axon. The release of neurotransmitters leads to the firing of an electrical impulse in the receiving (postsynaptic) neuron. Dopamine acts as a potent chemical messenger by binding to and activating cell-surface receptors in neighboring neurons. Researchers have found that individuals diagnosed with Parkinson's disease have one type of neuron that steadily degenerates, leading to less dopamine production; if there is a loss of neurons in the basal ganglia or if these vital dopaminergic neurons die, the levels of dopamine drop. This chemical imbalance is what results in the physical symptoms of tremor, stiffness, and slowness of movement.

The final punch endured by Ali—the last in the debates and perhaps the reason for Ali's knockout—are the genetic makeup and associated gene mutations that may be responsible for the underpinnings of Parkinson's disease. In 1997, mutations in the α-synuclein gene were found in individuals who suffered from Parkinson's disease.[56] Originally found in Italian and Greek families, the mutations have been discovered in both sporadic and genetic cases of Parkinson's, suggesting that α-synuclein, while a normal protein as discussed in Chapter 1, in this case can have a pathogenic effect. So the disease has strong underpinnings in a genetic or hereditary component.

This finding caught the attention of scores of researchers because, as you will recall from Chapter 1, the α-synuclein protein is usually present in synapses, the communication junctions that are formed between neurons. The protein is also highly enriched in the brain, accumulating particularly in the Lewy bodies, and is found in preponderance in severe traumatic brain injury and head trauma cases.[57] As α-synuclein protein increases, clusters will form in Lewy bodies that develop into deposits in the brain. Under the microscope, ubiquitin and α-synuclein proteins and α-synuclein-positive Lewy bodies appear to be spreading across all regions of the brain over time, a telltale sign that foreshadows a complication.

But studies of α-synuclein are just the beginning of our understanding of how genes influence neurodegeneration in Parkinson's disease, Alzheimer's disease, and dementia. Over the last two decades, population studies of more than two thousand individuals have been conducted to search for human genes associated with disorders like Parkinson's disease. Complete sequencing of DNA from each individual has uncovered numerous other genes—pink1, parkin, UCH1, LRRK2, and some two dozen more—that are known to promote Parkinson's disease after mutation.[58] And for a modest cost, individuals can pay private companies such as 23andMe, Human Longevity, and Ancestry to determine the sequence of DNA and look for genetic differences that could indicate a contribution to disease. Although similar to looking for a needle in a haystack, this approach gives clues to the cause of the disease.

But even with all of this progress and our four theories of cause, Parkinson's disease is still regarded as idiopathic, without a confirmed universal cause or mechanism of action. Beyond the four stated theories (the "punches" we associated with Mohammed Ali's case), additional hypotheses have recently been proposed. One idea postulates that the onset of Parkinson's is associated with the deposits of Lewy bodies, the aggregations of protein (α-synuclein protein) that develop inside nerve cells and also in the periphery, a theory supported by the finding that Lewy bodies also reside in the digestive system. Another hypothesis attributes the disease to long-term stress (a speculation made earlier in the case of Muhammad Ali's decline), which has been found to increase the vulnerability of dopaminergic neurons in the substantial nigra.[59] But while there are examples to support the role of stress hormones, whether they are involved in the onset or the worsening of movement problems has been difficult to determine.

A third theory highlights early clues witnessed in the periphery. Physicians have noticed that, like the appearance of constipation before the onset of motor symptoms, sensory problems such as a lack of smell arise in patients with Parkinson's many years before the onset of motor problems. It is estimated that 80–90 percent of Parkinson's patients experience a deficit in the ability to smell.[60]

Finally, an idea of current interest focuses on the location of α-synuclein proteins. We know that as α-synuclein protein increases, clusters will form in Lewy bodies, developing into deposits in the brain. But researchers have found that patients with a neurodegenerative disease sometimes have α-synuclein in the periphery or the enteric nervous system. The speculation here is that the vagus nerve, as a modulator of the brain and in its role connecting the brain and the gastrointestinal tract, plays an important role in Parkinson's disease. The hypothesis is that the disease originates in the periphery and is spread from the gut to the brain through the vagus nerve. It is an avant-garde idea without firm evidence, but one that is gaining some traction and attention.

Although we may never know what specifically caused or promoted Muhammad Ali's Parkinson's disease, considering medical histories like those of Mao and Ali allow us to find clues and consider new theories. More than two hundred years after Parkinson made his observations in London, it has been estimated that more than ten million people have been diagnosed with the disease. Although today we have an enormous amount of knowledge, there is still more to learn about medical problems like Parkinson's and other neurodegenerative diseases that affect the peripheral nervous system.

In the next chapter, we continue our investigation into the periphery and associated disorders with the aforementioned provocative hypotheses in mind. There is a good possibility that the role played by the glia—the collection of non-neuronal cells (cells that are not nerves) of the brain and nervous system—in the periphery may provide answers to important questions about both normal and abnormal states.

5

MORE THAN A GLUE

> A historical difficulty with studying glia has been the neurocentric view of the brain, implicit in the name of the discipline: neuroscience. Fortunately there is a growing appreciation of the importance of other cell types in the nervous system and their symbiotic relationship with neurons, with no single cell type now viewed as more important than the others.
>
> —NICOLA ALLEN AND BEN BARRES

AT ABOUT THE SAME TIME THAT JAMES PARKINSON was following his patients in London, the German scientist Theodor Schwann (1810–1882) was studying how muscle conduction was connected to neural tissues. Using advanced microscopes, Schwann observed the different cells that enveloped peripheral nerves. In 1837, Schwann proposed that all tissues were made of single cells and that cells are the fundamental unit of organization of all life. Arguably his most important discovery, however, was the glial cell that surrounded nerve fibers, later called the Schwann cell in his honor.[1]

Glia or glial cells were named from the Greek word for glue, *gloia*, since they physically touch and support neurons. They are non-neuronal cells found in both the peripheral nervous system and the central nervous system. Several glia are directly involved in producing the myelin sheath, the layer that forms around nerves and allows electrical impulses to transmit. Much of the spiral wrapping of the myelin sheath around axons happens after birth and after a child learns to hear, visualize, smell, taste, and balance. It is when a child is ready to walk that the rapid transmission of nerve impulses is necessary, for the development of proper gait along with proper speed requires nerve impulses. The process of myelination continues into adulthood and has been found to protect the axon from damage.

Beyond the important task of forming the myelin sheath, glial cells are responsible for getting rid of waste and an overload of neurotransmitters. They also interface with the immune system. Perhaps because they have so many roles, glial cells are much more numerous than neurons; in fact, most textbooks indicate a tenfold abundance of glia over neurons (a calculation that comes from the estimate of one hundred billion neurons and one trillion glial cells). Variations on this ratio have been reported, but the discrepancies reflect differences among brain regions.[2] Suffice it to say that the human body has a preponderance of glia or neuroglia cells. One of the most well known and important is the peripheral glia—the Schwann cell—responsible for many neuronal functions and for repair of the peripheral nervous system.

The Schwann Cell

The Schwann cell is best known for forming myelin, the aforementioned protective sheath covering nerve fibers and an insulating layer for axons that allows electrical impulses to transmit quickly along nerve cells in the peripheral nervous system (illustrated in Figure 5.1). These electrical signals are how nerve cells transmit information and facilitate movement between motor neurons and muscle fibers. If Schwann cell myelin is damaged, the impulses slow down and

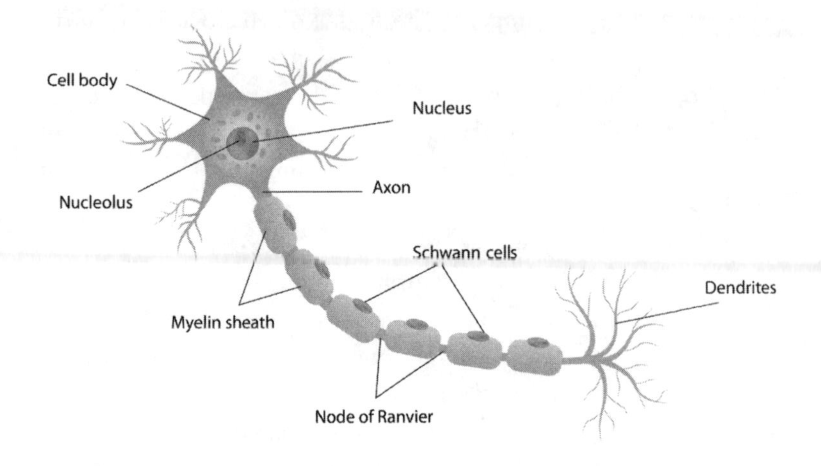

FIGURE 5.1 Schwann cells envelop axons in the periphery.

information is not passed along. Schwann cells also provide vital nutrients for neurons' survival and growth and produce growth factors. In short, they are indispensable.

To understand the relationship between Schwann cells and the peripheral nerves, it is important to remember that neurons consist of a cell body with two different kinds of branches: axons and dendrites. The axon—a nerve fiber—directly contacts the Schwann cells during development, triggering the production of myelin. We could consider peripheral nerves cables of electrical wires with an inner core—the axon—that are wrapped by the myelin sheet. Dendrites are appendages—branches—that arise from the body of the neuron and extend out to receive communications from other cells. Whereas axons carry impulses away from the cell body, dendrites carry them inward.

The generation of the myelin sheath around the axon is one of the most fascinating processes in biology. Only two cells, a neuron and the Schwann cell, are required to start the process of myelination, which continues after birth into adulthood. In the periphery, only axons with a diameter of one micrometer or more are myelinated. Myelin is not made by the axon but is derived as a continuous membrane extension of Schwann cells. To initiate the process of myelin formation, the Schwann cell and the neuronal axon must make direct contact.

The formation of myelin by Schwann cells is a remarkable undertaking, requiring two cells and a signal from the axon to the Schwann cell to initiate. As seen in Figure 5.2, the Schwann cell (SC) wraps around the axon (Ax) and makes many turns of the Schwann cell membranes. The myelin membranes are generated by the elongation and spiral-wrapping of the Schwann cell membranes around the axon. The membranes virtually glide over each other and then undergo compaction. It is a beautiful process.

The wrapping of Schwann cells around axons results in gaps that appear between adjacent Schwann cells (Figure 5.2). Louis-Antoine Ranvier (1835–1922), at the Paris Collège de France, was the first to observe this when he used a number of metal-containing dyes that revealed open gaps. Surprisingly, he found that the myelin coat was not present in the separation between Schwann cells. This finding turned out to be quite important because the gaps, now called nodes of Ranvier, function to speed nerve transmission. Nodes of Ranvier contribute to faster electric conduction by a phenomenon called saltatory transmission (from the Latin *salire*, "to leap"). The nerve impulse, an action potential, literally leaps from one node of Ranvier to the next.[3] Smaller axons that are unmyelinated do not have saltatory conduction and have slower action potentials, hence the decision to myelinate a nerve by Schwann cells has a profound effect on the periphery and the overall communication of cells throughout the peripheral and central nervous systems.

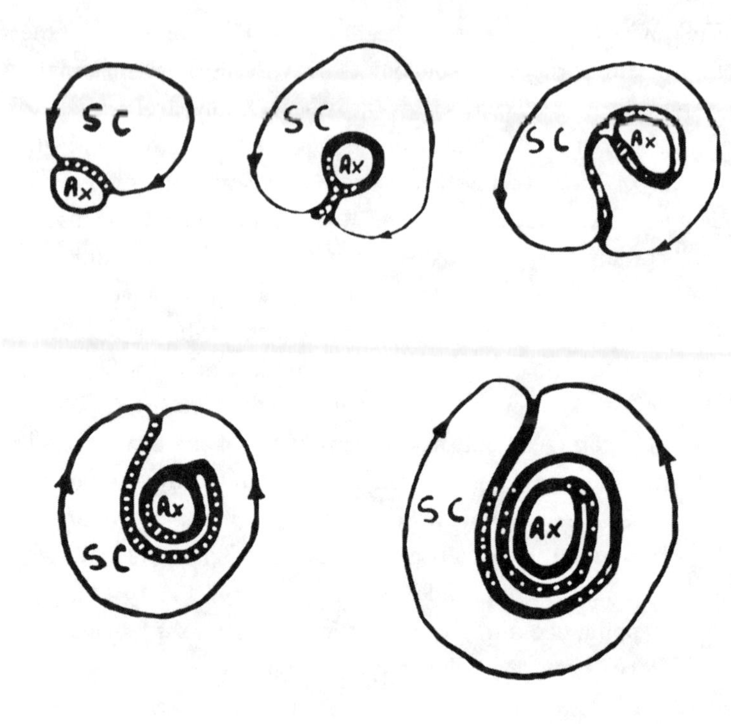

FIGURE 5.2 How a Schwann cells wraps around a peripheral axon. This classic drawing was made in 1954 by Betty Ben Geren, who used the electron microscope to visualize the process of myelination by Schwann cells (SC) wrapping around axons (Ax). The circular whorls of membrane become compacted into myelin. Anne Isabelle Boullerne, "The History of Myelin," *Experimental Neurology* 283 (2016): 431–45.

Myelin is strong and resilient compared to the rest of the body. It is such a hardy substance, so stable and durable, that the peripheral nerves of a prehistoric Iceman, recovered in 1991 after the man's demise some five thousand years ago, were so well preserved that the myelin sheaths could be detected by electron microscopy.[4] Myelinated nerve fibers in the brain and the skin were also preserved, as was the wrapping around nerve fibers, which was evident in regular layers. Even the nodes of Ranvier and myelin proteins were seen, whereas muscle tissue, blood, and connective tissues suffered from deterioration and could not be visualized. The evolutionary history of the gut was on display as well in the remarkably preserved remains of the Tyrolean Iceman; a paleo-proteomics analysis identified the *Helicobacter pylori* bacteria, a current member of the gut microbiome that is associated with inflammation in the stomach and gastric disease.[5]

Following Theodor Schwann's discoveries, Rudolf Carl Virchow (1821–1902) extended the cell theory by proposing that "every cell arises from another cell." He coined several cell biology terms that are in use widely today—*neuroglia*, *chromatin*, *parenchyma*—as well as the term *myelin*, derived from the Greek term for bone marrow (*myelos*).[6] Like Charcot, Virchow was a pathologist, passionate about studying cells that were abnormal or malformed, and a proponent of cellular pathology or histology, an area of research now represented in every medical center. On the faculty at the Universities of Berlin and Wurzburg, Virchow was known for his investigations of many diseases, from leukemia to hypothyroidism to typhus infection, and for studies of inflammation and amyloid proteins, which foreshadow the current interest in these processes during neurodegeneration. Virchow thought that all diseases could be traced to malfunctioning cells, not to tissues and organs. Virchow predicted the role of the myelin sheath in insulating the nerve and promoting the transmission of electricity. Later, an additional role was proposed: myelin provides energy to the nerve like a combustion engine. According to this idea, Schwann cells undergo a reaction in which chemical energy is liberated when in contact with nerve cells. Two present-day investigators, Klaus-Armin Nave and Jeffrey Rothstein, found that myelin indeed provides energetic support to the axon.[7] The dependence of the axon on myelin might explain why the axon eventually degenerates when there is loss of myelination.

In addition to changing nerve velocity, Schwann cells are also involved in the transmission of pain.[8] From the previous discussion, we learned that neuropathic pain is caused by injury or disease, resulting in damage to nerves or blood vessels and characterized by numbness and a lack of sensation or a sharp, stinging, and burning feeling. In the case of neuropathic pain, if Schwann cells disappear during nerve injury, axons may suffer losses from the lack of physical support and growth factors provided by Schwann cells. But whether the cells appear in the PNS or the CNS, when myelin sheaths are damaged, nerves do not conduct electrical impulses efficiently.

Information in the nervous system is transmitted as electrical signals that pass from nerve cells to muscle. Consider that all movements, whether one is stepping, swimming, or throwing a baseball, are facilitated by connections made between motor neurons and muscle fibers; the nerves and muscles work in tandem to initiate movement. Muscle fibers can either contract or relax as a result of contact with the motor neuron. A simplified view of a motor neuron extending its connections with muscle in the periphery is shown in Figure 5.3, in what is referred to as the neuromuscular junction, the place where synapses are formed and where the neurotransmitter acetylcholine is released.

Any movement, even a tremor, requires contact between motor neurons and the muscles that carry out the action. Movement—picking up a piece of paper, say, or walking to the kitchen—requires an exquisite coordination of speed and timing, nerve conduction, muscle contractions, and synapse formation. A dysfunction, an excess, or a paucity of movement due to paralysis or weakness of muscles signals a movement disorder. Hypokinesia, when movements are not as wide-ranging, and akinesia, a lack of movement, can be grouped together with bradykinesia, a symptom of Parkinson's disease.

The discovery of Schwann cells was fundamental because they helped distinguish the periphery from the brain, as the cells insulated

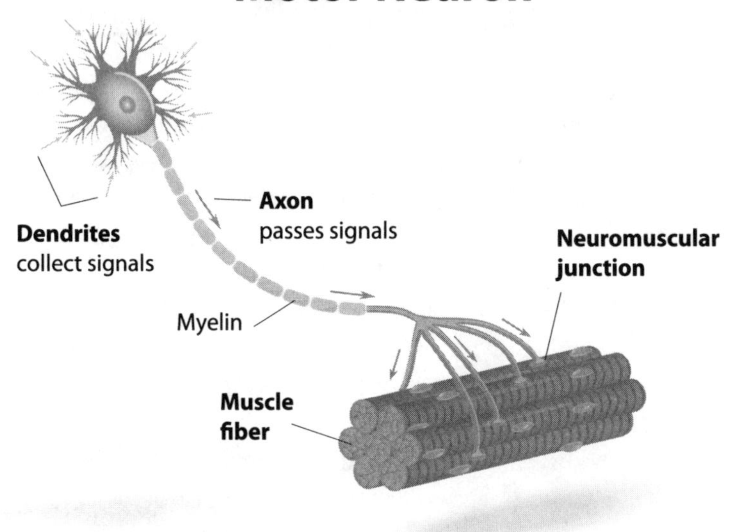

FIGURE 5.3 The contact between motor nerves and muscle is called the neuromuscular junction. Axons send signals to the periphery, and dendrites receive signals.

the axons of the nerve cells in the PNS and were not present in the brain or spinal cord. In the brain, the Schwann cells do not have the job of myelination; the oligodendrocyte cells do.

Oligodendrocyte Cells

In the early part of the twentieth century, the multitalented Pío del Río-Hortega (1882–1945) used sensitive silver stains to visualize the cells surrounding axons in the CNS. He called them oligodendroglia cells and argued that oligodendrocytes were a new support cell in the CNS. This theory went against his mentor's ideas; Santiago Ramón y

Cajal (1852–1934), the most famous neuroanatomist of all time, believed—incorrectly, as it turned out—that myelin substances were made and secreted from the axon.[9] We know today that myelin is not made by the axon; it is produced by glial cells in the periphery and the CNS. It was one of the few times Ramón y Cajal, a phenomenal neuroscientist, was incorrect. (As a matter of history, the methods used by the famous Italian neuroscientist Camillo Golgi, co-recipient of the 1906 Nobel Prize with Cajal, also did not detect oligodendroglia at that time.) It turned out that Río-Hortega was right.[10]

In the last chapter, I explained that the peripheral nerves originate from the neural crest, a layer of cells that eventually forms cartilage, bone, muscle, and the peripheral and enteric neurons. Early in development, Schwann cells, too, are derived from the neural crest, migrating long distances to form many structures in the periphery. In contrast, oligodendroglia come from a different embryonic site—the neuroectoderm—and are derived from progenitor cells that eventually generate the neurons and glia that make up the central nervous system. In human development, these cells appear in the embryo very early, eighteen days after fertilization.

In the CNS, oligodendrocytes insulate axons of nerve cells, like Schwann cells, but they make a different set of myelin proteins. Oligodendroglia do indeed produce myelin in white matter areas of the brain and spinal cord, wrapping hundreds of times around axons.[11] In contrast to a Schwann cell, which myelinates only one axon at a time, an oligodendrocyte can envelop multiple (up to thirty axons simultaneously), wrapping them up like an octopus (Figure 5.4).[12]

How can Schwann cells and oligodendrocytes differ in protein composition but have the same goal of myelination? I believe the best way to explain this is to use the musical analogy suggested by investigators at MIT, who translated into music protein structures arranged in dimers, sheets, helices, bundles, and many other configurations. The MIT investigators were able to do this since proteins have two- and three-dimensional shapes and arrangements that give different

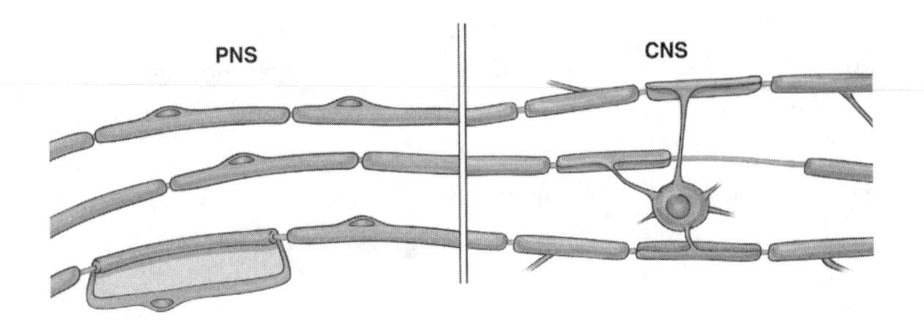

FIGURE 5.4 Differences in myelination between the peripheral nervous system (PNS) and the central nervous system (CNS).

conformations, arrangements that can be correlated to a different coding of vibrational frequencies.[13] Each protein gives a different signature based on twenty amino acids that can be converted into a completely different language or musical analog, and since each protein structure is distinctive, it can be represented by sine waves that make up musical notes. The analysis of proteins using musical scales is significant, since changes in protein sequence and structure underlie many diseases. Consider the prion protein we encountered in Chapter 3, which adopts different forms—a Jekyll and Hyde transformation—to cause transmissible neurodegenerative diseases. If we were to analyze the protein, we could see how it impacts injury, disorders, and possibly even regeneration.

But, important for this discussion, we have the Schwann cells in the PNS and the oligodendrocytes in the CNS. And here we must introduce yet another character, another glial cell type, the microglia. To do so, as the name suggests, we must peer closer to see a little more detail.

Microglia

Río-Hortega first identified microglia, largely from using silver dyes to study autopsied brain tissues of patients afflicted with neurological diseases.[14] These dyes gave him a greater degree of visibility to observe the fine structure of individual cells in autopsy material. His studies showed that cells derived from the middle layers enter the brain during the early stages of development, displaying bacteria-like morphology, and that they use blood vessels and white matter tracts to guide their invasion. Once established, microglia achieve a branched appearance, which is quite distinctive.

Microglia account for 10–15 percent of all cells found in the brain, and they play an essential role in brain disease.[15] Microglia serve as an active immune defense (they are the foot soldiers of the immune system), and they have a capacity to respond instantaneously to injury and inflammation, eliminating foreign objects without triggering an inflammatory response that might otherwise interfere with other neural functions.[16] Once activated, microglia are highly migratory cells that promote inflammation by producing cytokines, yet they can also respond to cytokines as part of the gut microbiome immune-signaling network. Microglia also serve a surveillance function after injury by keeping track of the local environment and by pruning and sculpting synapses or clearing debris as necessary, crucial tasks for the establishment of neural circuits.

Removing remnants of dying cells is another aspect of microglia's cleanup responsibility, its function as the body's cellular Grim Reaper. During development of the nervous system, microglia shape the functions of neurons, becoming an important process for building neural circuits. During early development, cell death is a predominant event, allowing correct connections to be made between immature neurons and their targets, such as muscle, skin, and the organs. Programmed cell death, or apoptosis, is an orderly form of cell suicide triggered by enzymes that digest proteins in the inside of cells (caspases). As a consequence, DNA becomes fragmented, and membranes swell and bubble, two classic features of cell death.

The origin of microglia has always been an ongoing debate. First it was thought that microglia might be derived from stem cells in the outer layer of the developing neural tube, or from a common progenitor cell with astrocytes. Microglia could also have come from the blood system in the periphery and specifically from progenitors from the yolk sac. In that regard, a recent study confirmed that microglia are derived from primitive blood-cell progenitors that arise in early embryonic stages. Cells from the yolk sac produce microglia at very early embryonic ages from macrophages appearing in the blood circulation that migrate to the brain.[17] Additional studies supported a middle mesoderm layer of the embryo as a source for microglia, again arguing that microglia may come from a population of cells called macrophages, which are derived from blood-cell precursors. These findings prove that microglia have connections in both the peripheral and the central nervous systems.

So microglia reside and act in the brain but emerge—originate—from the periphery. This fact indicates that the periphery plays an enormous role in shaping the cells of the nervous system, from glia to sensory neurons. One could say that the brain co-opted or stole microglia from macrophages located in the periphery. But for now, let's put our knowledge of Schwann cells, oligodendroglia, and microglia into practice and consider what occurs when there is a dysfunction in the axons.

Because long axons are more susceptible to damage, sensory and motor problems in the extremities are the most affected, which means the legs, hands, and arms experience atrophy and muscle weakness most often; damage is found mostly in the arms and legs in the peripheral nerves. Further, deterioration of the sensory axons may give rise to a decreased sensitivity to pain, heat, and touch in the feet and legs. This kind of impairment is often caused by genetic mutations, defects in specific genes that make proteins found in peripheral nerves. The disease is a hereditary motor and sensory neuropathy known as Charcot-Marie-Tooth (CMT), representing a large group of different inherited conditions that cause nerve damage.

Charcot-Marie-Tooth Disease

Pierre Marie, a French neurologist of the late nineteenth and early twentieth century, worked as an apprentice at the Pitié-Salpêtrière Hospital under Jean-Martin Charcot, the physician responsible for classifying and naming such disorders as ALS and Parkinson's diseases. In 1886, Marie and Charcot, together with a third collaborator, Howard Henry Tooth, published a paper, "Sur une forme particulière d'atrophie musculaire progressive, souvent familial, débutant par les peids et les jambes et atteignant plus tard les mains" (Concerning a special form of progressive muscular atrophy, often familial, starting in the feet and legs and later reaching the hands), in which they described a hereditary condition marked by a great deal of muscular degeneration. Marie also found a genetic form of cerebellar ataxia (derived from the Greek phrase *a taxis*, without order and coordination), which resulted in a loss of body movements and caused degeneration of the cerebellum.[18] This finding became known as Pierre Marie's ataxia.

Marie and Charcot's collaborator, Howard Henry Tooth, was a British neurologist who described progressive muscular degeneration in 1886. Tooth saw patients with severe muscle weakness and wasting in the legs and arms, many of whom had difficulty walking. The problem spread to the ankles and to the hands, where fine motor control was affected. And the lack of voluntary muscle activity hampered breathing, swallowing, and speech. He surmised that the disease was the result of a peripheral nerve problem and not a problem originating in muscle or the spinal cord. From the observations made by the threesome—Charcot, Marie, and Tooth—these peripheral nerve problems became formal symptoms of a new disorder known as Charcot-Marie-Tooth (CMT) disease.[19]

CMT is an inherited genetic disorder that damages peripheral nerves and causes muscle weakness and wasting in the extremities, lower legs, hands, feet, and forearms. The onset of the disease occurs during adolescence or in adulthood with a progressive weakness of muscles in the extremity, affecting foot, leg, and hand movements. Usually, the instability begins in the feet and ankles, with a foot that droops ("drop of the foot"). Because of paralysis or damage to the nerve, muscles that lift the foot near the ankle become weak. This causes friction between the toes and the ball of the foot, resulting in pain when walking. Tripping over a dropped foot or falling gives a hint to the clinician that CMT disorder might be present. As Tooth observed, when the disease progresses, the hands develop problems gripping doorknobs, buttoning clothes, and performing other fine-motor movements. Sensory neuron loss can trigger reactions in the skin, like hair loss and dry skin, and in more severe circumstances can lead to hearing loss and deafness. Damage to sensory nerves and loss of muscle can also bring about a lack of pain sensitivity or even neuropathic pain.

Not one disease, CMT now numbers over twenty-five related clinical disorders that affect one in 2,500 people in the United States and two to three million people worldwide. Although CMT has a genetic basis, different genes contribute to each condition. Because there are over ninety different genes associated with CMT, it has no simple solution, and it is difficult to know what causes each of the related disorders.

As an example, one disorder, now called Charcot foot, involves severe arthritis of the foot and ankle, often accompanied by bone fractures and diseases of the joint. As one can imagine, there were numerous theories for the cause of the disorder. In the 1880s, Charcot called the diagnosis arthropathies, which were derived from peripheral neuropathy or a loss of sensation. The French explanation was that the circulation and sympathetic nerves were somehow defective and caused the joint disease. On the other hand, Rudolf Virchow championed the German theory, arguing that trauma led to a loss of ability to sense the pain leading to the fractures (somewhat like the anti-NGF case). We still do not understand what causes the disorder.

Since CMT mutations have a great many targets, they can affect either nerve, glia, or myelin. A decrease in myelination decreases nerve conduction velocity and causes the axon to be vulnerable to degeneration, or an aberrant wrapping of Schwann cells around individual axons has direct consequences on nerve function. Further, demyelination can be triggered not only by genetic mutations but also by viral infection and inflammation. This makes it difficult to identify the guilty party.

Recall that movement results from an electrical signal from the brain to the spinal cord, and from there, the peripheral nerve passes the signal to the muscles. And axons, which can measure over a meter in length from the spinal cord, must send their signals in milliseconds. Many, but not all, axons are surrounded by myelin that insulates the electrical signals traveling through them, and myelinated axons and unmyelinated axons differ in their speed of electrical transmission. Defects in either the myelin or the axon cause a halt, a breakdown that results in progressive damage to the nerve cell, or if a mutation damages a peripheral nerve that connects the extremities with the spinal cord and brain, movement is affected. If the damage is in the axon, the strength of nerve conduction becomes weaker. Understanding the transport of materials and interactions of axons and Schwann cells will point the way for treatments for CMT.[20] Getting proteins and nutrients to the right place is one of the main goals of any cell, so following how unmutated cells and mutated cells function will assist with a treatment.

For instance, the most common form of CMT disease is known as CMT1. CMT1 is caused by a mutation of a myelin gene called peripheral myelin protein-22 (PMP22), which gives rise to a small protein embedded in the myelin membranes. In this case, the PMP22 gene is duplicated. Having an overabundance of this gene is damaging to Schwann cells, so one proposed way to treat this form of the disorder is to increase the number of Schwann cells to help keep the ratios of Schwann cells and axons intact.

Many CMT mutations are found in consanguineous families, where marriage occurs between relatives, because recessive gene mutations are inherited from ancestors. A case in point is a consanguineous CMT mutation, which was uncovered in India by pediatric neurologists. The symptoms displayed by a seven-year-old girl were similar to those present in other CMT cases, but the mutant gene turned out to be the problem. Her case report was striking in its description and detail:

> Since she was 7 years of age, her parents noticed that she had distal limb weakness initially in the lower limbs with frequent slipping of footwear. She then gradually developed upper limb distal weakness with difficulty in performing fine motor movements like writing and picking up small objects. . . . The child did not attend the follow-up outpatient clinic on a regular basis and only came back after three years when she was 10 years of age. During this period her weakness had progressively increased and she was now unable to get up from the floor. Her power had decreased in both upper and lower limbs.[21]

Both Schwann cells and axons are directly affected by this mutation. One idea is that one of many CMT mutations may be involved in sensing the size of axons by myelinating Schwann cells. This idea carries some weight because the body must regulate the number of myelin wraps around an axon, and the theory is consistent with a scaffold protein that acts as a bridge between Schwann cells and neurons. After all, inappropriate Schwann cell interactions with neighboring axons can lead to axonal damage, as shown in specific myelin mutations, such as in the PMP22 and connexin 32 genes. These myelin genes provide an explanation for how Schwann cells might promote demyelination in closely associated axonopathies. Therefore, I believe a focus on how Schwann cells contribute to Charcot-Marie-Tooth disorders will reveal the interconnections between neurons and Schwann cells in the periphery.

But what if there were a way to repair or regenerate damaged nerves or axons? Although axons cannot extend for more than one millimeter in the injured brain or spinal cord, peripheral nerves have the ability to extend well over one hundred millimeters. Schwann cells have also shown the ability to promote nerve regeneration.

Regeneration

The peripheral nervous system is capable of regeneration. The brain and the spinal cord are not. Why is this the case? Over the past century, there has been speculation, but there was not (and is not) a simple answer to this question. However, there are hypotheses.

One possible explanation is that the nerves in the central nervous system differed so much they lost the inherent capacity for regeneration. This idea turned out not to be entirely accurate. In a classic experiment published in 1981, neurologists Samuel David and Albert Aguayo, from McGill University, demonstrated that severed axons in the spinal cord actually grew long distances if they were associated with Schwann cells (in the periphery).[22] Over the last fifty years, Aguayo's research on nerve growth has revolutionized the study of regeneration. Aguayo, who received his medical training in Argentina at the University of Córdoba and spent his entire career in Canada as a professor of neurology and neurosurgery, was interested in how nerve fibers and glial cells interacted, both in the periphery and in the

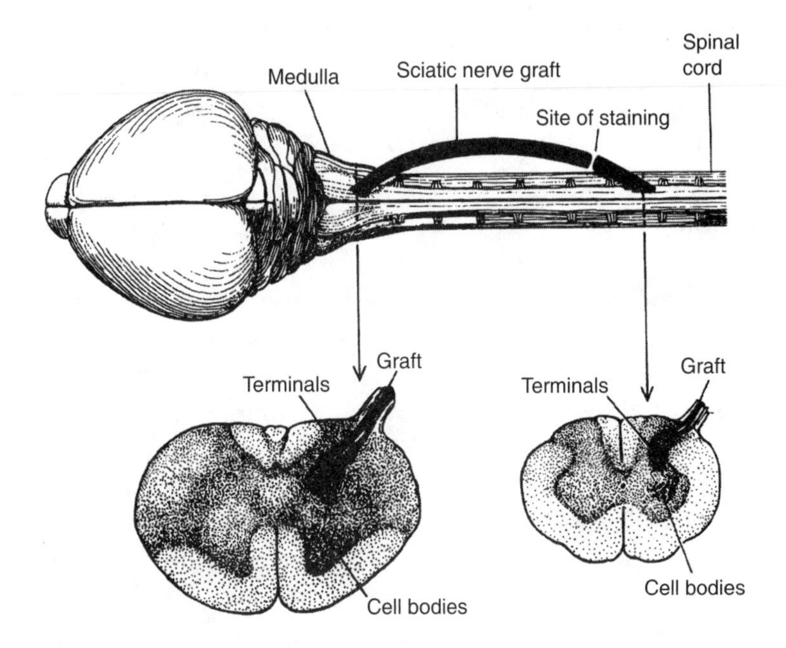

FIGURE 5.5 Lack of regeneration in the CNS can be overcome. CNS nerves send their fibers through a PNS graft containing Schwann cells (sciatic nerve graft). The top diagram shows a bridge made from the peripheral nerve graft placed between two locations in the spinal cord. The two cross-sections (bottom) show nerve fibers that course through each end of the graft. The nerve fibers are CNS neurons that have made their way through the Schwann cell bridge. Paul H. Patterson, "On the Importance of Being Inhibited, or Saying No to Growth Cones," *Neuron* 1 (1988): 261–67.

spinal cord. He noticed that glial cells in the right environment could be highly beneficial to nerve fiber growth, and his work suggested that the CNS environment was not appropriate for or permissive of regeneration. However, using a bridge made from a peripheral nerve graft placed between two locations in the spinal cord, Aguayo showed that Schwann cells from the periphery could promote the growth of axons in the spinal cord (Figure 5.5).[23] This simple experiment indicated there are ways to coax regeneration in the brain and spinal cord, despite the belief held by many in the field that spinal cord regeneration was not possible. Aguayo's findings were revelatory; the work changed how neuroscientists thought about the nervous system and stimulated a great deal of research on how to restore the functioning of nerves in cases of spinal cord injury or stroke.

Further, Aguayo's findings suggested that glia found in peripheral nerves (Schwann cells) were different from glia in the brain and spinal cord (oligodendrocytes). Other experiments showed that peripheral neurons could not grow in the presence of CNS glia. Based on these observations, it appears that the ability to regenerate after injury is dependent on the exact glia components in the environment of neurons.

Following David and Aguayo's experiments, two other hypotheses about regeneration have gained attention. The first theory points to a lack of trophic factors, such as NGF, which would prohibit regeneration. Since trophic or growth factors have been known to stimulate regeneration in the periphery, perhaps these factors are not present in the CNS, and this is the reason for a lack of regeneration. Paul Patterson (1900–65), a professor of biology at Caltech, speculated that "the lack of regeneration in the CNS is due to the absence of the necessary stimulatory or permissive substrates."[24] From 1960 to 1990, growth factors present in the CNS were investigated and debated, even though purifying the proteins in the nervous system was an arduous task. Eventually better ways of detection became available and showed that the same growth factors in the periphery (such as NGF, EGF, and insulin) are also present in the brain and spinal cord. So the inability to regenerate is not due simply to a lack of growth factors.

The second theory posits that a lack of growth may be due to a physical or chemical block, such as a "glial scar" barrier known to develop after nerve injury, or from an accumulation of inhibitory substances. In an initial attempt to repair the tissue damage, the local environment reacts by producing a network of proteins called the extracellular matrix. The net result is to produce a new physical environment that blocks nerve cells from regenerating. After a spinal cord injury, there is myelin debris and a number of cell types that are damaged—immune cells, astrocytes, and oligodendrocytes—which constitute the glial scar. This kind of environment is not conducive to regeneration. Before many others, Paul Patterson realized that "there was a growing list of potential cases of glia exerting an important influence on axon growth." In fact, "CNS glial membranes were involved in inhibiting axon extension," so Patterson was prophetic in promoting the concept of "the importance of being inhibited." Forces from glia prevented regeneration.[25]

To bring theory into practice, the most pressing question was, What cell types prevent regeneration (and in turn, which ones promote it)? It turns out that oligodendrocytes, the glial cells in the brain (not Schwann cells in the periphery), are a major cause of the problem. Axons assiduously avoid contact with oligodendrocytes because of specific proteins expressed on the surface of these cells; inhibitory cell surface proteins have a negative influence on regeneration in the spinal cord and the brain. One of the first proteins discovered by the Swiss scientist Martin Schwab was called "nogo," to convey that lack of growth or inhibition.[26] The theory that substances block regeneration has been verified with many other nogo-like proteins. Today we know that there are many inhibitory proteins expressed by oligodendrocytes and myelin that block regeneration. Not many of these proteins are found in Schwann cells.

Now this is a puzzle. So far, the theories that have been posited have not been proven. Perhaps we should put the previously stated question another way: Why do myelinating cells in the periphery promote regeneration but myelinating cells in the brain do not?

One possible explanation is the need to shape the morphology of neurons through inhibiting growth and the targets that are innervated. Another way of patterning cells in the brain is by a process called "tiling," in which there is a self-avoidance and restriction of cells in a particular zone. (We will reconsider this problem in Chapter 8, on plasticity.) Further, because myelin continues to be made into late adulthood, it has long been thought that changes and remodeling of myelin may occur from experience and neuronal activity. Measurements to visualize white matter myelin have shown there is a great deal of variability as a result of experience and exposure to different activities, which is the base definition of plasticity.[27] Neuronal activity can, in turn, regulate changes in myelin. Therefore, myelin plasticity may go hand in hand with neuronal plasticity. This notion comes from detection of changes in myelin thickness with an increased ability to perform motor tasks, such as piano playing and figure skating. On the other hand, other paradigms, such as stress, social isolation, and sensory deprivation, have a negative effect on myelination.

Keeping in mind our earlier discussion of myelin, the continuous formation of new myelin, the variability in its thickness, and the changes in the lengths between the nodes of Ranvier—all intriguing properties that suggest myelin is much more than a covering, or glue, of axons. The detection of gradients and changes in myelin profiles has stimulated the search to find out if myelin remodeling could be responsible for the acquisition of new motor tasks, new learning, and increased cognition.

Clearing Out Waste

Besides their role in the production of myelin and protection and support of neurons, glia are responsible for other duties that are important to highlight, notably in the lymphatic system. Often called the body's sewage system, the lymphatics are a part of the body's circulation and also a part of the immune system. The lymphatic system is a fluid hydraulic system that is essential for the inner workings of the periphery. Lined with endothelial cells and other non-neuronal cells, lymphatic capillaries carry lymph, a colorless fluid, to the bloodstream and pick up fluid that leads into tissues. It is a drainage system, but it also protects the body against infection by making white blood cells—T and B lymphocytes.

Earlier we encountered microglia that originate in the blood circulation and find their way to the brain. Apparently, the brain lacks a true lymphatic system. However, it does have fluid made by brain ventricles circulating through it: cerebrospinal fluid (CSF). CSF constantly circulates, much like lymph, but the movement is mediated by glial cells and requires astrocyte water channels found in the lining of the brain's fluid spaces to clear the fluid. These specialized channels help move the CSF efficiently and reduce the resistance of the traveling fluid. An interaction between CSF and the blood system exists and involves active exchange with the periphery.[28]

To distinguish CSF from the lymphatic system, researchers have designated the "glymphatic" system to acknowledge the contribution of glial cells to this process. The glymphatic system collects the sewage in CSF from cells in the body and delivers it to the circulation. Using a variety of tracers to mark lymphatic vessels and the glymphatic pathway, one clearly sees that CSF fluid can be cleared through specific anatomical routes that follow white matter tracts and other drainage veins in the circulation. Tracers injected into the brain are easily followed in this circulation.

It appears that the primary role of the glymphatic pathway is clearance of waste and other metabolites. But there is also a role for distributing nutrients, such as glucose. When the glymphatic system is damaged, as in the case of traumatic brain injury, CSF flow is impacted, which can be catastrophic to memory and motor abilities. In fact, stroke is another malady that is affected by the flow of fluids in the glymphatic pathway.

Clearing fluid in the periphery through the lymphatics has another provocative function. Maiken Nedergaard and Steven Goldman from the Universities of Rochester and Copenhagen sought an answer to the question, Why are lymphatic vessels important in the brain?[29] Aware of its drainage properties, Nedergaard went on to show that β-amyloid is a major molecule that is cleared from the brain. What's more, she found that the clearance occurs most often during sleep. As much as two-thirds of the β-amyloid is cleared by the lymphatic system. This finding shows an amazing connection between the function of sleep and the periphery. Due to the high metabolic activity of neural cells, it is essential that waste products of neural metabolism are quickly and efficiently removed from the brain interstitial space. Several degradation products of cellular activity, such as β-amyloid oligomers and amyloid proteins, can alter synaptic transmission and calcium concentrations and trigger irreversible neuronal injury.[30] In other words, during sleep, the lymphatic system clears out waste. Supporting this growing appreciation of the importance of the sleep state in relieving the body of the burden of metabolic catabolites, researchers found that the rodent brain removes waste better during sleep or anesthesia than during the awake state.[31] This finding leads to an inference that the role of sleep is to switch into a state that facilitates the clearance of degradation products of neural activity that accumulate during wakefulness.

The role of sleep in assisting in the removal of waste material has important implications, especially with what we have learned so far about health and disorders associated with the periphery. Recall that sleep disorders are one of the symptoms of Parkinson's disease. Patients with peripheral neuromuscular diseases often present with difficulties sleeping. Issues with sleep-disordered breathing has been seen in CMT, and sleep apnea occurs in CMT1. Lastly, in patients with other CMT diseases, an impairment in peripheral nerves is frequently seen, including problems with the vocal cords and the diaphragm, along with poor sleep quality due to fragmentation, leading to reductions of REM sleep.[32] So bucking the traditional view that the cellular mechanisms of sleep are neuronal, I believe that glial brain cells are actively engaged in an important role during sleep. The glia's ability to respond to chemical signals further suggests that they could regulate neuronal activity, notably to detoxify the brain during sleep.

Another implication is that the glymphatic system likely becomes less active with age. Poor glymphatic circulation or clearing of waste and toxins may lead to Parkinson's and Alzheimer's disease, likely because of the buildup of toxic proteins such as τ- and β-amyloid in an age-related decline.

Just as we have seen earlier in the chapter, glial cells have the capacity for cell division and activation, to undergo change and respond to the environment.[33] Rejuvenation of the lymphatic vessels and glymphatic activity may be a feasible approach to correcting deficiencies in the circulation. Activity-dependent changes in glia have been proposed for learning and memory and for motor skills. These ideas suggest that glial cells and myelin could be responsible as a cell biological mechanism for plasticity.

This chapter's discussion of glial cells was an important piece of the puzzle. We learned how our glial cells assist in development and respond to and care for the environment. We will discuss this potential more in the upcoming chapters, but for now, we know that glial cells are much more than glue.

6

A LACK OF TEARS

In the course of the past 10 years four children have appeared at Babies Hospital with symptoms so puzzling as to defy exact diagnosis yet so similar as to constitute a clinical entity. A fifth patient with the same condition has been studied at the New York Hospital. The common features are: 1. Undue reaction to mild anxiety characterized by excessive sweating and salivation, red blotching of the skin and transient by marked arterial hypertension, and 2. constantly diminished production of tears. Search among textbooks and current literature has failed to reveal a description of a syndrome exactly corresponding to the one described.

—CONRAD M. RILEY, RICHARD L. DAY, DAVID MCL. GREELEY, AND WILLIAM S. LANGFORD (1949)

CONRAD M. RILEY (1914–2005) and Richard Day (1915–89) were baffled. They had never seen anything like it. The young children, ranging in age from five to eight years old, showed dramatic swings in blood pressure, frequent drooling, excessive sweating, and, most surprisingly, an inability to produce tears. As pediatricians and as professors at the College of Physicians and Surgeons at Columbia University, Riley and Day were the first to observe and document a puzzling, rare genetic disease afflicting Jewish children of Eastern European descent. For most children and adults, tears are a normal product of strong emotions and have been documented as such for centuries. Yet even while exhibiting varying degrees of excessive anxiety or drama, the children whom Riley and Day observed lacked the ability to produce tears.

At first, the pediatricians could not find any other reports of these symptoms in the medical literature, especially in children. Riley and Day detailed their observations of one patient in particular in their seminal publication in 1949:

> This patient was seen first at the age of 6 yrs., 5 mos., because of attacks of vomiting since the age of 6 mos. During the 29 months following her first visit, or until the age of 8 yrs., 10 mos., she was admitted to the Babies Hospital four times for treatment and study and between hospital visits she was seen regularly in the outpatient department . . . Until about the age of 10 years the patient perspired so profusely when she went to bed that her bedclothes had to be changed about 1.5 hours after retiring. She would also perspire and drool when excited and alarmed.[1]

Riley and Day found through observations of patients such as the above that the syndrome that now bears their names, also referred to as familial dysautonomia, results in a lack of tears and a pronounced sweating response. In a healthy individual, tears are produced in a sweat gland known as the lacrimal gland, located inside the upper lid of each eye. Although the inability to form tears may be disconcerting in an emotional way and in interpersonal circumstances, it is also a medical problem, since tears function to protect the cornea from drying out. And a lack of perspiration—or profuse perspiration as in the patient Riley and Day referred to—is also a personal and medical concern, since either under- or overproduction of fluids from sweat glands makes it difficult to maintain or regulate body temperature.

Familial Dysautonomia

The term *dysautonomia* refers to mistakes in controlling involuntary actions, such as the regulation of tears, sweat, and blood pressure.[2] *Familial* refers to mutation of a specific inherited gene; researchers now know that familial dysautonomia is an autosomal recessive disease, meaning that two copies of the defective gene must be present in the offspring for the disease to develop.

Familial dysautonomia is a disorder with a genetic connection. Historically, the mutant gene that accounts for Riley-Day syndrome can be traced back to the 1500s to a small Jewish population in Eastern Europe.[3] Catherine the Great, the empress of Russia, decreed in 1791 that the Jewish population be restricted to the Pale of Settlement, an area that now includes Poland and Ukraine. As a result of forced segregation, the carrier rate of the familial dysautonomia mutation in European Jews was very high, affecting one in thirty-two people in the Jewish population who came from regions in Poland and Ukraine. During Riley and Day's studies in the 1940s, cases mirrored the high numbers of Hasidic and Orthodox Jewish families who had moved to New York, although today the number of familial dysautonomia births is one in ten thousand in North America. Cases of familial dysautonomia primarily follow the pattern of migration of the Jewish population from Europe to North America, although some recent cases have been found in Spanish-speaking areas such as Mexico, where families were not aware of a Jewish heritage, and across South America, Great Britain, and South Africa.

Familial dysautonomia is an early-developmental disorder, first detected in newborn children, that often worsens over time. The origin of the disorder is a disruption in the development of the peripheral nervous system that affects neurons in the sensory and sympathetic systems. When in Chapter 1 I likened the sympathetic and parasympathetic systems to first and second violins in an orchestral string section, I noted that each component must balance the other. The sympathetic nervous system sends the powerful "go" signal to promote fast action in the body; it is responsible for stimulation. The parasympathetic system does the opposite: it slows the response to the original impulse, bringing conditions back to normal. These two systems are responsible for shifting blood supply to the muscles, digestion of food, regulation of immune responses, wound healing, and insulin secretion. An imbalance of the systems or an overregulation of the sympathetic system—our initiating "go" signal—can lead to extreme symptoms in the body, such as overabundance of perspiration.[4] A similar sensory role is played by the baroreceptors, specialized neurons in the arteries that respond to the stretching of blood vessels, which maintain blood pressure, by providing a feedback loop. Elevated blood pressure causes the heart rate to decrease, whereas decreased blood pressure causes the heart rate to increase. This is once again a product of the fine balancing that our bodies go through to support our health and is a prime example of the peripheral nervous system's communication.

But in any disorder, the perfect balance is disrupted. Symptoms of familial dysautonomia first appear during infancy. Signs include poor muscle tone, poor growth, and feeding difficulties, as well as the symptoms noted earlier: a lack of tears, difficulty maintaining body temperature, and swings in blood pressure. Other symptoms can include social withdrawal, uncommunicative behavior, difficulties with elimination (either diarrhea or constipation), and sometimes a slowing of reflexes or a loss of bodily movement control, a condition called ataxia. In this state, balance or proprioception is poor, so walking becomes uncoordinated and imbalanced.[5]

If we were to look inside the body at each muscle, we would see spindles, stretch receptors that detect changes in the length of the muscle and convey the length information to the central nervous system via nerve fibers. They are a crucial component of the system of motor control, which, when operating normally, protects against injury caused by overstretching. But in Riley-Day syndrome, the muscle spindles do not respond or communicate, which is what leads to a decrease in balance and a loss of fine motor control.[6]

If we were similarly to look into the body to see what was afoot with sweat glands that caused increased perspiration, we would find, surprisingly, that the patients' sweat glands appear normal in number and size. How can this be? In an effort to find the answer, investigators measured the conduction of peripheral nerves. In these tests, electrodes are placed on the skin of peripheral sensory or motor nerves. After sensory and motor nerves are stimulated, electrical activity in the muscle is measured. This measurement can confirm if a peripheral nerve is directly involved and if the nerve is damaged or degenerated. Researchers can also perform a biopsy on a small piece of skin from the leg or the back (a punch biopsy) and examine the sample under a microscope.[7] Although the sweat glands of the patients with familial dysautonomia at first appeared to be normal, researchers found that the glands had a significant loss of peripheral nerves and sparse connections with the remaining nerves. Research also revealed unhealthy and empty-looking Schwann cells. This telltale sign of empty Schwann cell sheaths is another confirmation of nerve loss in the periphery. Perhaps that is why glial cells play such a key regulatory role.

Like those with Parkinson's disease and Charcot-Marie-Tooth disorders, patients with familial dysautonomia also have sleep difficulties, most often breathing disruptions while sleeping. In fact, nearly 90 percent of patients with familial dysautonomia have this symptom.[8] Often this is due to an obstruction of the pharynx or upper airway, which becomes either partially or fully blocked, a condition called obstructive sleep apnea syndrome that can produce disordered breathing similar to what we saw in the CMT case in the previous chapter.[9] These issues are so serious that sleep is one of the most dangerous activities for familial dysautonomia patients, as long pauses in breathing can bring on sleep-related sudden death or build on other medical issues.[10]

A dramatic example is a 2018 case of a fifty-four-year-old man suffering from familial dysautonomia. One evening, his breathing became so labored that he was admitted to the emergency room in an unconscious state. After he was revived, doctors discovered that he had food in his lung, which blocked his air supply. In this man's case, and in others, familial dysautonomia created a loss of peripheral nerves along the entire gastrointestinal tract. As a result, movement of food down into the stomach was blocked and that food became stuck in the esophagus. In this case, food in the patient's digestive system was aspirated into his lungs, which prevented him from breathing.[11]

Although there are so many complications ushered in by familial dysautonomia, and it is difficult to follow which event happens first and which emerges next, there is one certainty: the periphery is where the action is. As I have described, the autonomic system, which includes the sympathetic, parasympathetic, and enteric nerves, is at the center of this malady. When the autonomic system goes awry, many physical changes arise, and the issues begin to build on one another, adding to symptoms we have previously discussed, such as the increase of perspiration and uncontrolled blood pressure.[12] In the case of increased blood pressure, sometimes an unexpected part of the periphery comes into play: the kidneys.

Although the kidneys are not usually covered in textbooks about the nervous system, the kidney is an important organ when it comes to blood pressure. A kidney's main function is to act as an elaborate blood-filtering system that removes waste products and excess fluid from the body. It must be an important job, because we evolved two kidneys to operate as filtration systems. Each kidney is made up of about one million tiny filtering units called nephrons. Each nephron has a filter, called the glomerulus, that allows fluid, waste products, and nutrients to pass through but prevents blood cells and large molecules, mostly proteins, from passing. The filtered fluid then passes through the tubule, which sends needed minerals back to the bloodstream and removes wastes. The final product becomes urine, which flows from the kidneys to the bladder through tubes called ureters. For most people, two kidneys usually filter nearly 150 quarts of blood per day to produce about one to two quarts of urine, composed of water, wastes, and extra fluid.

Like a Rube Goldberg contraption containing a labyrinth of stop-and-go choices, the kidneys have intricate checks and balances to regulate blood pressure and fluid balance. When blood levels become low, the kidneys respond by releasing an enzyme called renin to convert a protein from the liver to a hormone called angiotensin I. This response from the kidneys sets off a perpetual chain of reactions. The lungs provide a second enzyme to generate angiotensin II from angiotensin I. These hormones cause blood vessels to constrict and blood pressure to increase. Angiotensin II also stimulates the release of another hormone called aldosterone in the adrenal glands, which sit above the kidneys.[13] Aldosterone causes the kidneys to raise blood volume, blood pressure, and sodium levels in the blood to restore the balance of salts and fluids. If the renin-angiotensin signals become too active, high blood pressure results. In other words, signals from the sympathetic system send directions for the contraction of blood vessels, causing a reduction of blood flow in the kidneys; the kidneys receive input from both the sympathetic and the parasympathetic nervous system to maintain balance.

The kidneys can also change blood pressure by increasing or decreasing the amount of urine that is produced. When the kidneys make more urine, the amount of blood that fills the arteries and veins decreases, and blood pressure is lowered. If the kidneys make less urine, the amount of blood increases in the arteries and veins, which increases blood pressure.

In healthy patients, nerve fibers are found all along arteries that course through each kidney. A biopsy of the kidneys in dysautonomia patients showed a marked reduction in the number of sympathetic nerves that innervate blood vessels in the kidneys. These relationships suggest that patients with familial dysautonomia have a peripheral abnormality that makes them more vulnerable to changes in blood pressure.[14]

Another symptom of dysautonomia is fainting, or syncope, a temporary loss of consciousness usually related to insufficient blood to the brain. Syncope, seriously studied only since 1986, has been documented for hundreds of years. Recall from Chapter 4 that William Shakespeare used the condition in his most famous plays, including *Julius Caesar,* when the ambitious and charismatic leader faints and falls after being offered the royal crown of Rome, resulting in a joyful and excited "falling sickness." Centuries after Shakespeare's death, in 1616, we now realize that "passing out" is due to a temporary lack of blood flow, usually from low blood pressure, resulting in insufficient blood flow to the brain that leads to dizziness, fainting, and blurred vision.[15] This malady can also occur as a result of changes in the rate of the heartbeat, drastic changes in body metabolism—from lack of oxygen or low sugar levels—and even changes in the environment; standing too long in a hot room, for example, can make one feel faint, and astronauts returning to earth after experiencing zero gravity are susceptible to vasovagal syncope.

How is syncope detected? One way is to perform a tilt-table test. In this test, an individual first lies flat on a table (mimicking a fainting episode) before slowly moving to an upright position. Doing so will determine if there is a decrease in blood pressure, the hallmark of orthostatic hypotension, which occurs when parts of the peripheral nervous system undergo degeneration and cannot compensate for changes in posture. A normal result of the test is for the blood pressure to stay stable. If the blood pressure drops and stays low, the heart rate cannot adapt to the changing positions, often resulting in lightheadedness and blurred vision.

Much of what we have learned about familial dysautonomia, how the body responds to the disorder, as well as how the body functions overall to balance systems, comes from research that occurred after Riley and Day's observations. One study in particular gave us insight into what happens to nerves in this disease and established the role of the periphery in Riley-Day syndrome. The study was led by a neurologist whom we met when we discussed experiments he led on CNS regeneration: Albert Aguayo.

The Return of Aguayo

In 1969, Albert Aguayo became interested in dysautonomia. When he was an assistant professor at McGill University, Aguayo came across a man who had been seen previously by Dr. Conrad Riley as a three-year-old patient. The history of this patient was similar to that of others Riley and Day had seen in that he also suffered from lack of tears and perspired excessively, but the description also pointed to some new potential symptoms, an aspiration of food:

> Feeding problems were present from the first weeks of life. By the age of 3 years, he had been hospitalized 13 times for recurrent chest infections, corneal ulcerations, hyperpyrexia [extreme temperature] of unknown etiology, hypotensive episodes, vomiting leading to dehydration and, on one occasion, seizures.[16]

Because he understood how the periphery works, Aguayo suspected that the problems may have arisen from changes in the peripheral nerve and nerve conduction, especially after the patient showed slower measures of nerve conduction velocity. To look for deficits in the nerve itself, Dr. Aguayo took a biopsy of the patient's sural nerve, a major sensory nerve located in the leg that gives sensation to the skin of the feet and ankle. After removing and fixing the nerve, fibers were teased apart using fine forceps. The fibers, approximately 1.5 centimeters in length, were then cut in much smaller pieces and examined under a microscope that could magnify the nerves a thousand times. Not satisfied, Aguayo also examined many fibers under the electron microscope, which can magnify the nerves ten thousand times. Each sample was cut into dozens of ultrathin sections to obtain a series of non-overlapping pictures. The microscope pictures revealed that there were fewer unmyelinated fibers than one would expect to find, and the results were published in a 1971 *Neurology* article titled "Peripheral Nerve Abnormalities in the Riley-Day Syndrome."[17]

The loss of unmyelinated and large myelinated nerve fibers implied that there was a problem during the development of the nerves. As previously discussed, since Schwann cells fully encircle the nerve fibers and provide trophic support to neurons, the presence of empty Schwann cells was an indication of a serious problem. Like the skin biopsy, it is a sign that can account for loss of nerves in dysautonomia.

Although only one patient was examined, Aguayo's analysis got to the bottom of the problem. The defects in these nerves suggested a problem in the survival and ultimate fate of these neurons in early development, an idea originally proposed by two legendary embryologists, Rita Levi-Montalcini and Victor Hamburger, in the 1940s. Using clever detective work, Aguayo found that unmyelinated nerves were specifically affected in the sural nerve. As it turns out, the nerves were classified as mechanoreceptors in the C-class. C-class peripheral nerves have a small diameter and are specialized to detect pleasant touch, as well as deep pain. These particular sensory nerves are found in hairy skin and respond to slow and light stroking. And they are found in the peripheral nervous system.

Puzzle Pieces

Although familial dysautonomia is caused by malfunctions of the autonomic nervous system, we can consider the disorder a triple threat, as it affects the autonomic, sensory, and motor neurons. We can see this breadth of issues in the symptoms and functions we have outlined; the loss of tears and excessive perspiration, change in blood pressure, kidney function, syncope, and body and skin temperature control are signs of problems in the autonomic nervous system, and an inflammation of the joints in the foot or ankle gives rise to a loss of sensation or neuropathy in the extremities that involve sensory and motor neurons.[18]

Although the diseases have different origins and causes, Parkinson's disease and familial dysautonomia have considerable overlaps.[19] Recall that the mutations in α-synuclein became a factor in Parkinson's disease and that aggregation of α-synuclein is also associated with dysautonomia. As discussed in the previous chapter, during development, peripheral neurons increase the length of their axons to make connections with other sensory neurons and muscle fibers and to turn on genes and move molecules and nutrients. If there is difficulty in making these connections—a case of abnormal transport—neuropathy and CMT diseases can result, which can affect sensory, motor, enteric, and autonomic neurons. Now the pieces of the puzzle are starting to come together.

But there are a number of other ideas to investigate as well. Could there be similarities between familial dysautonomia, Parkinson's disease, and other degenerative disorders? Perhaps a genetic connection? It turns out that there could be a defect in nerve growth factor (NGF). We have discussed how NGF is required for the development and maturation of the peripheral nervous system, particularly the sensory and sympathetic branches of the autonomic system that require it for survival. A lack of NGF in the periphery causes serious problems in sensory detection of pain, temperature, and respiration, and the number of myelinated axons can also be compromised.[20]

Recall that the discovery of NGF came from an effort to understand what controls nerve growth in the periphery. This result led to the next question: Do inductive or diffusible substances control the peripheral nervous system, like sensory and sympathetic nerves? The answer lies in secreted trophic factors, as they are imperative for proper survival and circuitry of the nervous system.[21] Early competition during development sculpts the way the nervous system forms and determines which neurons live and which undergo death (Figure 6.1). The target of these neurons is a determining factor. Since the targets of peripheral nerves produce only minute amounts of NGF, nerve cells that make the correct connections with their targets and encounter NGF are the victors.[22] Cells that are mistargeted to the wrong location or do not respond to NGF cannot compete and lose the battle. They undergo cell suicide and programmed cell death.

Figure 6.1 depicts the survival scenario. First, newly born neurons in the periphery grow their axons. Then they seek to make connections in the periphery to the skin, muscles, glands, and all the organs. As noted previously, the nerve cells that make the right connections to the target survive by contacting trophic factors like NGF and avoiding apoptosis, programmed cell death. Many nerve cells do not survive this process, so there is substantial cell death. In this case, the process is part of a healthy ecosystem, despite the death of half of the neurons that are born during development. The process ensures proper connections by regulating the size and connectivity of neuronal populations for a functioning system and then prunes away an overabundance of cells.

So if a lack of connections with targets leads nerve cells to not encounter NGF and die as a result, could NGF be defective in a degenerative disease like familial dysautonomia? Researchers found that

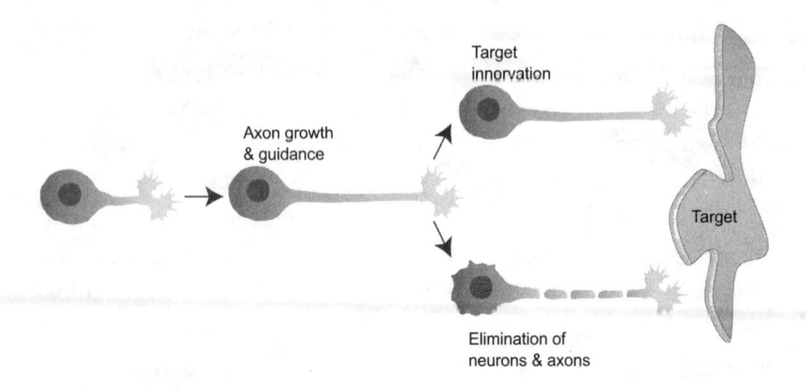

FIGURE 6.1 Development of the peripheral nervous system. Immature neurons on the left are influenced by trophic factors that help them migrate to the appropriate targets in the periphery.

surprisingly, NGF was not the gene that directly causes Riley-Day syndrome. Not all hypotheses turn out to be correct.

However, in a landmark discovery, the familial dysautonomia (FD) gene was first identified on human chromosome 9 in 1993 by a group led by James Gusella at Massachusetts General Hospital.[23] By mapping the genes of dysautonomia families, researchers identified the FD gene as ELP1 (elongator protein 1), which has replaced the complicated name IKBKAP (inhibitor of kappa b kinase complex-associated protein gene). The ELP1 gene is responsible for familial dysautonomia.

Unknown until recently, ELP1 controls myriad signals in neurons. It is a complicated gene with many responsibilities, including the production of proteins, the development of sensory and sympathetic neurons, regulation of blood pressure, and feeding and swallowing. The ELP1 gene determines the fate of nerve cells by correctly processing genetic information in the nucleus, namely the flow of instructions from DNA to RNA. The ELP1 protein is part of a large complex involved in the transcription and processing of gene products. It is responsible for making the correct RNA species and, conversely, the proteins made from the RNA.

Researchers found that individuals with familial dysautonomia have two copies of an ELP1 mutation, which results in a reduced amount of the elongator protein and produces errors in processing ELP1 mRNA, as well as many other mRNAs.[24] Further, investigators found that these mutations have a reduced number of neurons in the dorsal root ganglion and sympathetic ganglion, and the resulting neurodegeneration of peripheral nerves can lead to widespread organ dysfunction and increased mortality. Lastly, the lack of ELP1, like a lack of NGF, leads to the selective degradation of sensory and autonomic neurons and affects the enteric neurons, which could explain the many gastrointestinal issues associated with familial dysautonomia.[25] That one gene mutation is responsible for a number of functions of the autonomic nervous system: heart and respiratory rates, digestion, and urination.

To take the thousand-foot view, we should consider the ELP1 gene an essential gene. It has a fundamental biological role in the development of humans and other organisms, as it is also evolutionarily conserved in many species, including fruit flies, fish, and worms. In the absence of the ELP1, many connections in the body between nerves and their targets are not made correctly, ultimately leading to death.

All of this information supports Rita Levi-Montalcini's theory that target-derived trophic factors ensure connections between immature neurons and their target during development. And, fitting two puzzle pieces together, sympathetic and sensory neurons in ELP1-deficient mice showed a marked decrease in transport of NGF.

We have the lifelong work of several researchers to thank for these findings. Dr. Felicia Axelrod, a professor who specialized in pediatrics and established the Dysautonomia Treatment and Evaluation Center at the New York University School of Medicine in 1970, after identifying many dysautonomia patients, established a special clinic for treatment and research that has studied over 650 individual cases. Of the many ways to diagnose individuals with familial dysautonomia, Axelrod and others use genetic tests to determine whether individuals carry the mutated ELP1 gene. Axelrod's research has been followed by Dr. Horacio Kaufmann at New York University and Dr. David Goldstein at the National Institute of Neurological Disorders and Stroke, investigators who have concentrated their efforts on pursuing the complications that arise from the autonomic nervous system.[26] Kaufmann and Goldstein observed early losses of neurons in patients' autonomic nervous system, before degeneration of neurons in the basal ganglia. Perhaps we can consider this work an early sign of things to come.

Similar Symptoms

Through the efforts of numerous investigators and a number of pharmaceutical and clinical approaches, there has recently been impressive progress in the understanding and treatment of degenerative diseases, many with symptoms similar to those of familial dysautonomia, including spinal muscular atrophy (SMA) and other hereditary sensory and autonomic neuropathy (HSAN) disorders.[27] SMA is a deadly genetic condition that weakens the nerves in the spinal cord, impairing simple functions like breathing, swallowing, and mobility. Without the SMN gene, motor neurons die, and muscles become weaker.

In a remarkable series of efforts by Adrian Krainer, a researcher at Cold Spring Harbor Laboratories, processing defects by the SMA gene have been corrected using genetic engineering to fix a splicing error in the mutated SMA gene.[28] In a similar manner using a short piece of DNA that is an antisense oligonucleotide against the site of the mutation in the ELP1 gene, Krainer corrected the defect in familial dysautonomia.[29] In these landmark procedures, it was found that the positive effects from the oligonucleotide (in the periphery) compensated for deficiencies in the CNS.

Furthermore, HSANs, a family of disorders involving sensory and autonomic issues, are characterized by mostly peripheral problems, mainly the inability to feel pain and temperature. These neurodevelopmental disorders are caused by a mutation in single genes, diagnosed by measuring sensory and autonomic function. Although each of the genetic disorders in sensory and autonomic nerves represents mutations in different genes, they have symptoms that will be familiar.

HSAN I patients have a mutated enzyme that processes membrane lipids abnormally. Those afflicted have a sensory loss of pain and temperature sensation in their lower limbs.[30] HSAN II patients exhibit severe feeding and breathing problems, and HSAN type III is the new classification of the original Riley-Day diagnosis, or familial dysautonomia.[31]

Patients with HSAN IV, also referred to as congenital insensitivity to pain (CIPA), do not feel pain, changes in temperature, or touch, due to a loss of NGF-dependent peripheral neurons.[32] Other familiar symptoms arise in HSAN IV, such as a loss of the ability to perspire, which makes sense, as sympathetic nerves in the autonomic nervous system stimulate glands in the skin to secrete sweat. Researchers have also found mutations in the gene for TrkA, the receptor for NGF.[33] Recall from our previous discussions on pain and itch that the NGF-TrkA pair is responsible for sensing and communicating pain, itch, and inflammation; TrkA receptors are involved in pain signals and its ligand, NGF, supports the survival of peripheral neurons during embryonic development. Mutations in the receptor gene cause NGF-dependent peripheral neurons to undergo cell death. Lastly, in HSAN V, mutations have been found in the NGF gene. Since NGF is a peripheral pain detector, the bodies of people carrying an NGF mutation have difficulty with injury detection, tissue damage, and increases in inflammation.[34]

While HSAN hereditary disorders are rare, research suggests that many of the symptoms are seen in more prevalent disorders, maladies we have discussed throughout this book. It is conceivable that the symptoms of familial dysautonomia may eventually be reversed by similar approaches, as the symptoms all appear to point to issues that either begin or are foreshadowed in the periphery.

As we have seen, the PNS controls many different organs that regulate digestion, sweating, touch, and heart rate, malfunctions of which can create a snowballing of medical issues. Beyond the disorders we have already discussed, there is considerable connection with some neurological and psychiatric diseases. In the next chapter, we will consider one of them—autism, a modern ailment that is rooted in problems of social behavior and anxiety. Although autism is a different disorder from familial dysautonomia and Parkinson's disease, it is astonishing how many health-related problems are shared by all three.

7

THE POWER OF TOUCH

> The five senses are how each of us understands everything that isn't us. Sight, sound, smell, taste, and touch are the five ways—the *only* five ways—that the universe can communicate with us . . . But what if your senses don't work normally? . . . Then your experience of the world around you will be radically different from everyone else's, maybe even painfully so.
>
> —TEMPLE GRANDIN

ACCORDING TO RECENT ESTIMATES by the Centers for Disease Control and Prevention, one in sixty children is affected by autism.[1] A developmental disorder, autism impairs individuals' ability to communicate and interact with others, evinced by the lack of non-verbal forms of social interaction, such as making eye contact and frowning, smiling, or other changes in facial gestures in response to another person. A person with autism may have a difficult time understanding what others are thinking and may also take attempts at humor literally. Social communication is also characterized by language where grammar, syntax, and logic play key roles. Young children with autism often have delays in speech and language skills. And in both children and adults, individuals with autism show repetitive behaviors like lining up toys, fixating on objects, and sometimes even rocking back and forth. Such behaviors may also reflect an attempt to tune out sensory stimulation.[2] In fact, a hallmark of autism is hypersensitivity to noise, touch, and other senses.

Three of these interrelated symptoms of autism—a lack of communication skills, a lack of social interaction, and persistent repetitive stereotyped patterns of behavior—have resulted in numerous studies to determine if family genetics contributes to brain differences found in the disorder. Indeed, researchers have discovered that many genes and brain circuits are implicated in autistic traits seen from the earliest age.[3] But based on the writings of Temple Grandin, a spokesperson on autism, and on a breadth of research on the senses in the periphery, I believe it is time to consider the connections between the periphery and the disorder.

The periphery, specifically related to the sense of touch, can explain the problems encountered in autism. As discussed in previous chapters, nerves found in the skin are very sensitive to touch, temperature, vibration, and such sensations as itching or tickling. Our previous discussions and studies of Parkinson's disease, familial dysautonomia, and other disorders have indicated that events in the periphery can either be responsible for disorders or provide a warning of a disorder to come. And there are many similarities between autism and the previously discussed diseases.

But before I continue my case for the periphery, it is important to mention an outlying finding for autism not shown in other disorders we have discussed: autism has a skewed sex ratio. Three times more boys than girls are diagnosed with autism.[4] Why? Researchers are still working on this puzzle, but there are theories that the sex ratio may reflect differences in the diagnosis rather than in the occurrence. For instance, girls may show fewer repetitive responses and have more socially acceptable ways to express themselves that make autism more difficult to detect. Girls may also receive a diagnosis later in life, during adolescence, when another behavior related to autism becomes more prevalent: anxiety.

A response to difficulties in communication and behavior, anxiety often appears during adolescence for a host of reasons, including peer pressure, the intensity of schoolwork, and an increase in the production of hormones. Unsurprisingly, stress is highly associated with anxiety and depression. (Here is where we begin to tread on more familiar territory.) Stress is our evolutionary fight-or-flight reaction, a sympathetic response. Recall from Chapter 2 that glucocorticoid steroids (stress hormones) made by the adrenal gland increase and decrease in response to stress, which can lead to feelings of anxiety and depression and affect learning and thinking. To understand how anxiety has been manifested in autism, I take you back to the historic work of physician-scientists in nineteenth-century Europe, and to a familiar physician.

Observations

Jean-Martin Charcot, whom we met before, was the most famous French physician of his time, the first to name multiple sclerosis and fifteen other conditions, including Parkinson's disease and Charcot-Marie-Tooth disease. In our previous discussion, I noted that he had excellent observational skills that he brought to his clinical work and his drawings. But there is more to his story.

Charcot had a flair for the dramatic, which perhaps was one of the reasons he studied psychiatry. When he lectured for colleagues and students, he offered flamboyant demonstrations, describing patients suffering from anxiety and hysteria. And in a lesser-known twist of history, Charcot's studies in neuroses and the basis of hysteria, as well as hypnosis as a therapy, attracted a soon-to-be-famous student, Sigmund Freud (1856–1939).[5]

As a twenty-nine-year-old medical student, Freud went to Paris in 1885 for a short sabbatical, leaving his medical studies at the Vienna General Hospital to work under the guidance of the already famous physician Jean-Martin Charcot. After this brief visit, he and Charcot became fast friends and close colleagues. It was in Paris where Freud learned the powers of clinical observation, leading to the new field of psychoanalysis. There was no shortage of disorders to observe at the Pitié-Salpêtrière Hospital, from hysteria to anxiety, and behaviors including disjointed speech and fainting spells. Even nervous impulses, such as movements independent of voluntary action, were seen on a daily basis.

In 1885, many of the behaviors Charcot and Freud studied may have been related to autism, but the term *autism* (Greek for "self") was not coined until 1908, by Eugen Bleuler, a Swiss psychiatrist interested in schizophrenia, although autism was not formally identified as a disorder until 1943.[6] Freud's clinical work on nervous disorders laid the foundation for future investigations into autism, including studies by Bleuler. Aware of Freud's description of autoeroticism and theories of narcissism during early childhood, Bleuler extended this line of thinking by proposing that autism is characterized by individuals who were self-absorbed, introverted, and disengaged with the outside world. The term he used—*autism*—foresaw the development of child psychiatry and the future prevalence of the disorder.

The clinical diagnosis of autism in 1943 is actually traced to Leo Kanner (1894–1981), a physician at Johns Hopkins University, who described a group of children with unique features who "have come into the world with innate inability to form the usual, biologically provided affective contact with people, just as other children come into the world with innate physical or intellectual handicaps."[7]

Implicit in Leo Kanner's idea was that autism is a condition determined by family genetics, a sizable shift from Bleuler's earlier theories of narcissism. Kanner based his hypothesis on his clinical observations of a group of eleven children. These children often displayed temper tantrums and communication handicaps.[8] One timid young female patient displayed obsessive and repetitive behavior and had difficulty expressing herself with correct syntax and words, even though she showed sufficient progress in reading, writing, and spelling. Another patient was afraid of moving objects and mechanical sounds. She could not relate to the other children or other people and had difficulty with language, particularly with pronouns and negative statements.

Aware of Freud's work, Kanner first pursued the psychoanalytic idea that poor parenting was the root cause of autistic behaviors, but then he realized that there were other reasons for the disorder that could not be blamed on poor child rearing. It was true that the patients had few relationships with family and friends, and they had behavior and language issues; however, they also showed abnormal reactions to pain and to hot and cold temperatures.[9] The sensory reactions were enough to make Kanner consider that there may be biological reasons for the disorder.

Another physician at that time, Hans Asperger (1906–80), based in Austria, observed children who exhibited atypical problems of repetitive behaviors and lack of social interactions.[10] However, the children he studied were high functioning. Asperger reported the conditions of boys between the ages of seven and ten and noted that they exhibited competence in mathematics and possessed greater facility with language and developed more highly elaborate vocabularies than the patients Kanner observed. Asperger realized that this syndrome might be a separate disorder.

Although Kanner and Asperger worked on different continents during World War II, each arrived at the same diagnosis of autism during the 1940s. Leo Kanner's published article "Autistic Disturbances of Affective Contact" appeared at about the same time in 1943 as Asperger's postgraduate thesis. Kanner received much more attention than Asperger, perhaps partly because Asperger's thesis was published in German and remained obscure for many years. But there was a bit of controversy; although Kanner grew up in Germany and clearly could read Asperger's thesis, he did not acknowledge Asperger's theory. Today, clinical researchers believe that the subjects Asperger and Kanner studied were at different ends of the same, broad spectrum and refer to the disorder as autism spectrum disorder (ASD).

Moving on to a different controversy, questions about Asperger's character have recently come to light. In an unexpected revelation, Edith Sheffer's *Asperger's Children: The Origins of Autism in Nazi Vienna* implicates Hans Asperger as a participant in child euthanasia during the Third Reich.[11] The book asserts that although an opponent of Hitler's National Socialist Party, Asperger became increasingly involved in Nazi euthanasia programs in the 1940s. A second publication by Herwig Czech, "Hans Asperger, National Socialism, and 'Race Hygiene' in Nazi-Era Vienna," also highlights Asperger's poor treatment of his patients in Austria.[12] It is difficult to look at the physician's studies in the same way after reading this information.

A second controversy surrounding Asperger concerns the disorder classification. For decades, there was disagreement about whether to define the condition as a type of autism, since the patients diagnosed with Asperger's were higher functioning than other patients with autism. Some clinicians felt that the definition was too broad and questioned the validity of an Asperger's diagnosis. One key difference these clinicians questioned is found in diagnosis: Was there a developmental delay observed within the first three years? Patients diagnosed with Asperger's did not show significant cognitive delay. Other criteria used intelligence and language skills, which are difficult to measure and can reflect social and racial bias.

Why was there so much concern about combining or differentiating these similar disorders? The reason was not scientific. It boiled down to medical coding and billing by insurance companies and services offered based on diagnosis. Payment for treatment is dependent on defining the diagnostic criteria using the official DSM-5 manual in psychiatry and the International Classification of

Diseases.[13] Services provided (and paid for) were also dependent on the category.

I suppose we could say that the second controversy has been laid to rest, but since it is intricately affixed to medical coding and the insurance industry, one may never know. Regardless of how these disorders are classified and categorized, we need to learn more about how and why there is such a disturbance in a body's equilibrium, and we need to continue our investigation of how the periphery is involved.

Genes and the Environment

Much of autism research focuses on the circuitry and the actions in the brain, and a mapping of the brain regions that contribute to autistic behavior. The hope is that patterns of brain activity at an early age will help predict autistic and repetitive behaviors that arise during infancy. In other words, a substantial amount of research is focusing on faulty signals in the brain (sound familiar?). Other researchers believe there could be an underlying problem that extends throughout the body. In other words, a disease develops as a result of one's inheritance and environmental influences, or, put simply, the key lies in the genes and environment behind behavior.

To investigate the latter idea, researchers have found that small alterations in identified genes can contribute to Parkinson's and CMT conditions, and for autism the DNA sequencing of subjects has resulted in the classification of many genes, more than eight hundred, that display mutations. The Simons Foundation Simplex Collection assembled a bank of genetic samples, using specimens from families. One family set included a child with autism, their siblings, and at least two unaffected parents. This strategy has identified many hereditary family connections in autistic offspring. Similar genetic analysis has led to a greater understanding of the genetic ties involved in attention deficit disorder and schizophrenia as well as major depression, bipolar disease, and post-traumatic stress disorder. Curiously, autism and schizophrenia share many genetic patterns, in addition to sensory processing problems.[14]

Many gene mutations affect the formation of synapses, including neurotransmission and the structure of synapses, which can lead to faulty connections between nerve cells that give rise to electrical and chemical signals. However, unlike in the case of familial dysautonomia, autism does not have a single gene with which it is associated—it has hundreds. While this makes correlations more complicated, one could say that autism bears some similarity to the impact of synaptic plasticity used to explain many neurological conditions.

It is also complicated—daunting, really—to determine the connectivity of an autistic brain because defects are not located in just one circuit or in only one specific part of the brain, such as the basal ganglia. A mechanism is needed to explain how genes in the brain are directly involved with specific circuits and behaviors of autistic individuals. We must first understand what causes repetitive behaviors, communication problems, and lack of social interaction. Given many possibilities, how do we pare down all the potential explanations for the involvement of genes and the environment? Let's look a bit more at a symptom mentioned a few times already in this chapter: anxiety.

Anxiety and Stress

Recent research has shown that young children who are hyper-responsive to sound, touch, or visual cues have a greater risk of developing anxiety later. For this reason, investigators are looking at the strong relationship between the sensory symptoms and the social behavior of autistic individuals.[15] To measure anxiety levels in children, investigators at the University of Pittsburgh developed a set of questions known as SCARED: Screen for Child Anxiety Related Emotional Disorders.

The children were given a battery of questions and then asked to rate, with simple response marks, if the answer was *very true* (++), *somewhat true* (+), or *hardly ever true* (–) anytime in the last three months. Sample questions are listed below.[16] The questions are also applicable to the type of behaviors that stimulated Charcot's and Freud's interest in anxiety over a century ago.

Statements to Respond

1. When I feel frightened, it is hard to breathe
2. I get headaches when I am in school
3. I don't like to be with people I don't know well
4. I get scared if I sleep away from home
5. When I get frightened, I feel like passing out
6. People tell me I look nervous
7. I feel nervous with people I don't know well
8. I get stomach aches at school
9. I worry about sleeping alone
10. I worry about being as good as other kids
11. I worry about going to school
12. When I get frightened, my heart beats faster
13. I have nightmares about something bad happening to me
14. I worry about things working out for me
15. I get really frightened for no reason at all
16. When I get frightened, I swear a lot
17. I am a worrier
18. I am afraid to be alone in the house
19. It is hard for me to talk with people I don't know well
20. People tell me that I worry too much
21. I don't like to be away from my family
22. I am afraid of having anxiety or panic attacks
23. I feel shy with people I don't know well

24. I worry about what is going to happen in the future
25. When I get frightened, I feel like throwing up
26. I worry about how well I do things
27. I am scared to go to school
28. I worry about things that have already happened
29. When I get frightened, I feel dizzy

The patients (children), ages nine to nineteen, and their parents took this test. Interestingly, the parents' scores correlated more often with those of adolescents (twelve years of age or older) than with those of younger children (nine to twelve years), suggesting that some traits may be passed on from generation to generation and that social or biological (hormonal) influences may also be a factor. Putting some of the social and hormonal considerations in perspective, scientists found that there was enough overall consistency in the final scores to make this a valid and reliable test of anxiety disorder, as well as a test for autism.[17] The questions in this test are reminiscent of an autistic boy described in the book *The Curious Incident of the Dog in the Night-Time*, by Mark Haddon. As an example of anxiety, the young boy is quoted as having reactions not so different from those addressed in the SCARED test:

> I do not like people shouting at me. It makes me scared they are going to hit or touch me and I do not know what is going to happen. . . . Talking to strangers is not something I usually do. I do not like talking to strangers. . . . I do not like strangers because I do not like people I have never met before. They are hard to understand. . . . It takes me a long time to get used to people I do not know.[18]

Today, clinical scientists look at responses to quizzes, alongside analysis of behaviors and genetic tests, to identify the disorder and potential underlying familial causes of illnesses. Unsurprisingly, large-scale genetic studies have found a treasure trove of genes responsible for neuronal development, behavior, and way of living, and over 100 "autism" genes have been discovered from an analysis of over eleven thousand individuals.[19] But remember, there isn't one gene responsible for the disorder. This abundance of data has promise but is overwhelming. Imagine a hundred million puzzle pieces dumped on your floor.

Thomas Bourgeron, a professor studying the genetics and traits of autism at the Institut Pasteur in Paris, has argued that although a list of genes is valuable, there is a greater need for a general theory that can account for all the traits of autism.[20] This unifying theory requires an extensive study of all the genes and cells, the anatomy of circuits, and how they relate to behavior. We must find answers to questions such as, Is there a trigger for this disorder? What of effects of the environment on autism? In the case of Parkinson's, studied for centuries, what is the initial cause of the disease?

We do not yet have that unifying theory, nor have we been able to answer these questions for complex disorders like autism and Parkinson's disease. A recent editorial by Paul Nurse, a former president of Rockefeller University and a Nobel laureate, stated the issue with beautiful simplicity: "It is easy to accumulate large mountains of data, but it is not as easy to come up with a working theory. More concepts, theories, and imagination are needed to generate explanations."[21] For example, brain imaging is a powerful way to find the differences between Alzheimer's and other dementias, but insights into why they are different are difficult to glean. Nurse agrees with Bourgeron: we need more ideas. To which I add, we need strategies that are outside the norm.

One idea is to observe the impact of a stressful environment. The following historical event illustrates how social interaction and communication problems can arise. In Romania during the twentieth century, over one hundred thousand children were raised in state-run institutions under crowded conditions without sensory stimulation.[22] These children suffered from a lack of caregiving and social interaction and were also victims of neglect and abuse. Unsurprisingly, as a result, many of the children had social problems and a susceptibility to anxiety or stress.[23] Some of the children developed repetitive behaviors and cognitive issues that persisted into adulthood.[24] More than 10 percent of the children were later diagnosed as autistic. These outcomes illustrate the strong influence of social and sensory deprivation on disposition and personality. Yet they do not explain the many cases of patients who did not experience the same kind of deprivation during their development.

So while the brain, genetics, and the environment may have a correlation with the development of autism, we need to continue to search for answers. It may just be that the peripheral nervous system can help to explain autistic sensory perception and response, especially to pain.[25]

A Sensory Connection

Temple Grandin, the autism advocate and animal behavior scientist, has been vocal about the possibility that sensory deprivation is an explanation for autistic behavior. Grandin, who herself is autistic, observed that many individuals diagnosed with autism disorder have altered or exceptional sensory experiences involving touch, temperature, and pain. She aptly documented how tactile stimuli influenced her own reaction:

> Small itches and scratches that most people ignored were torture. A scratchy petticoat was like sand paper rubbing my skin raw. Hair washing was also awful. When mother scrubbed my hair, my scalp hurt. I also had problems with

> adapting to new types of clothes. It took several days for me to stop feeling a new type of clothing on my body; whereas a normal person adapts to the change from pants to a dress in five minutes. New underwear causes great discomfort, and I have to wash it before I can wear it. Many people with autism prefer soft cotton against the skin. I also liked long pants, because I disliked the feeling of my legs touching each other.[26]

As Temple Grandin's words illustrate, she was one of many individuals with autism who was sensitive to different kinds of touch. This conclusion has been substantiated by many studies, which show that abnormal or hypersensitive sensory reactions occur in more than 90 percent of autistic individuals.[27] A greater reaction or hyperreactivity to touch comes from specialized nerve endings in the skin that can sense when something even as light as a feather touches the skin. For that reason, researchers hypothesize that hyperexcitability within the peripheral sensory system could explain why autism subjects have strong negative reactions to touch.

After all, touch is a form of social communication that plays a critical role during childhood in bonding, attachment, intimacy, learning, and memory. It is one of the first functional modalities in an infant and has a central role in the emotional growth of children. An article by Simon Baron-Cohen, a recently knighted professor in Cambridge, offers the opinion that there is a strong influence of "abnormal touch"—a disinterest in touch or texture—in autism and during development.[28] This finding, based on abundant human evidence, is very similar to what Grandin described. Baron-Cohen also observed a clear connection between sensory behavior and repetitive behaviors such as rocking back and forth, nail biting, and hand twisting; patterns of sensory and repetitive behaviors often go hand in hand.[29] Tactile perception and hypersensitivity have been shown to have a role in socialization, development, and also repetitive behaviors.[30] Further, researchers have found that increased sensitivity is a mirror of abnormal sensory abilities and is more likely during the early years which could result in sensory problems or alterations in the ability to see or taste or touch normally.[31]

Children with autism often have difficulties controlling the gross and fine motor movements required for handwriting. Small groups of children with autism, between the ages of eight and thirteen, have been tested for handwriting skills, specifically their ability to write several sentences legibly, using correct spacing and alignment. The subjects were asked to copy the words "The brown jumped lazy fox quick dogs over" and were directed to make sure the letters were the same size. An example of the results of a handwriting test are shown in Figure 7.1.

As you can see from the example, children with autism had more difficulty with legibility and with producing the right size, spacing, and alignment.[32] From issues forming letters correctly to irregular spacing between words, many children could not fully control and regulate handwriting movements, which led to poor handwriting quality.[33] While a number of the studies focus on problems in the brain (cognition) to account for these types of motor movement issues, problems could also be attributed to the skin and the periphery, as this complex skill requires motor planning and perceptual and tactile sensitivities.

As the largest sensory organ, the skin has been described as a "social organ" by virtue of its role in social perception and communication.[34] Our skin is divided into hairy and non-hairy, or glabrous, types. Hairy skin is found over most of the body. Sometimes therapists work with children diagnosed with autism using affective touch, or a gentle, slow touch on the hairy skin on an arm to convey social support and to increase social bonding. Glabrous skin, found on the lips, the palms, and the soles of the feet, helps us understand subtle tactile details like texture and shape. Although skin is a protective barrier, there are many elaborate structures inside the skin designed

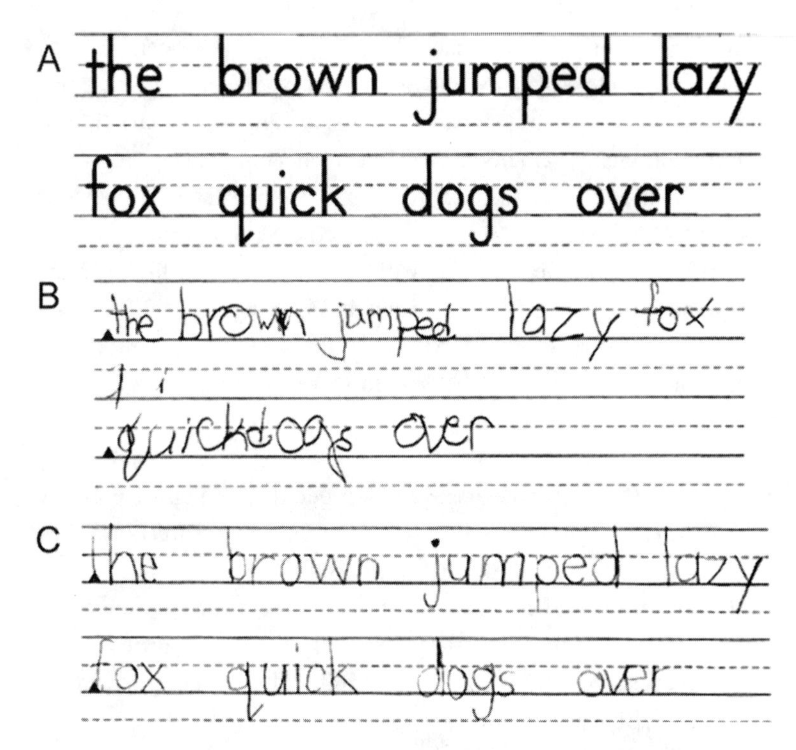

FIGURE 7.1 Handwriting changes in autism. (A) is the sentence that was provided; (B) is handwriting from an autism subject with a low PANESS and Wechsler score; (C) is handwriting from a higher-scoring ASD subject. Subjects were tested beforehand for motor skills (PANESS, Physical and Neurological Exam for Motor Signs) and on the Wechsler Intelligence for Children. T. Tavassoli and S. Baron-Cohen, "Taste Identification in Adults with Autism Spectrum Conditions," *Journal of Autism and Developmental Disorders* 42 (2012): 1419–24.

to detect touch and pressure. These sensory neurons can be further divided on the basis of their nerve conduction velocities.

The wide assortment of receptors found in the skin allows us to sense touch, respond to mechanoreceptors in the environment, and grasp an object. As we have seen in evolution, touch is required for survival, obtaining food, and even for pair bonding or reproduction.

Accordingly, our skin has adapted to be perceptive to gentle touch and is essential for social exchange. It has also adapted to be sensitive to harmful and warning signals. There is a tremendous variation in duration, intensity, velocity, and frequency of contact.[35]

Touch (and pain) are sensed by neurons in the peripheral nervous system through sensitive receptors. A tactile sense of touch requires specialized sensors to assist us with fine motor tasks like playing the piano and typing on the computer. Mechanoreceptors, such as Merkel's cells, are specialized to perceive form and texture, and Meissner's corpuscles, found in the very top layer of skin, sense different levels of vibration and touch under the skin with high sensitivity.[36] Ruffini's corpuscles can sense stretching and movement, and other receptors in the skin can even record temperature, alongside pressure and texture.[37] All these receptors are classified by the threshold—low or high—and whether they are myelinated.

To communicate information about the state of the body and environment—It's too hot! That's prickly!—sensory neurons transmit messages to the central nervous system. They extend their axons to the spinal cord and toward targets in the periphery. Sensory receptors like thermoreceptors measure changes in temperature and possess the ability to sense heat (capsaicin) or cold (menthol), as mentioned in Chapter 3, and nociceptors or free nerve endings sense pain and are sensitive to perturbations in the skin. And although we often think of sensory neurons in the skin as responding to our body and the environment, these receptors are also found in the heart, muscle, and bone.

Lastly, a sense of location and the detection of balance are handled by proprioceptive neurons. These neurons are found in the periphery in close proximity to other sensory nerves, and they have the ability to locate where our body is and our spatial orientation and posture. This ability allows us to determine our movements, speed, and the position and direction of our arms, legs, and torso. In his book *The Man*

Who Mistook His Wife for a Hat, Oliver Sacks described what happens when an individual loses proprioception.[38] One can lose hand coordination, have difficulty standing, and even feel like one's body is not their own. Some physicians have argued that proprioception should be a sixth sense, joining touch, smell, taste, sight, and hearing. Such a designation would certainly bring more awareness of connections between the periphery and medical issues.

The Power of Touch

David Ginty, a distinguished professor of neurobiology at Harvard University, has extensively studied the origins and location of sensory nerves in skin. His laboratory made a contemporary observation about the power of touch, one that makes important connections between sensory nerves and genetics. Mice carrying different autism spectrum gene mutations were analyzed for touch responses and thresholds. Several of the mutations, including a gene called MECP2, which controls the expression of hundreds, even thousands, of genes, were characterized in severe forms of autism, such as Rett syndrome, a severe disorder that mainly affects girls.

In a paper published in 2016 by Ginty's group, mice carrying Rett mutations in MECP2 and other human autism genes produced greater sensitivity to touch.[39] This hypersensitivity was readily observed alongside greater anxiety and deficits in social interaction in animal behavior tests. Ginty's group then traced the different responses to individual groups of peripheral neurons from earlier work on developmental history and anatomical location. When human autism gene mutations were studied in mice, researchers used a sensitive assay to detect if the subjects displayed a startle response to an air puff. The mice that had the human autism gene mutation were very sensitive to just a light air puff to the skin, giving rise to a startle reaction. Researchers also found that the mice had difficulty discriminating between novel textures, and contrary to most expectations about brain involvement, sensory neuron terminals were hypersensitive to even the gentlest touch. Even a challenge with a noise or an air puff prompted a peripheral response. This sequence of events and their results suggest that the defects could originate in the peripheral nervous system, not only in the brain.[40]

Studies of tactile communication in animals and humans show that touch often has an effect on social communication, emotion, and attachment, as well as learning and memory.[41] This is the case for humans with and without autism. But for individuals with autism, every sense—touch, vision, smell, and taste—has been observed to create a sensitive reaction.[42] For instance, the smell of fear through sweat or a startle often produces a recoiling response in autistic adults.[43] Such reactions can arise in response to biological chemicals in the environment with aversive effects, or it is possible that the development of these kinds of behaviors can be attributed to a lack of social communication. However, although some people believe autism conditions should be defined solely in terms of social symptoms, I disagree. There are studies that support the argument for a malfunction of sensory processing; blinded studies carried out on children with autism, ages eight to twelve, found impaired processing of textures and vibrations underlying their inability to filter sensory information appropriately.[44] And observations of sensory over-reactivity and abnormalities in tactile responses in autistic mouse models have led to the proposal that anxiety reactions and mechanosensory neurons originate in the periphery.

One idea to address touch sensitivity is to reduce sensory neuron dysfunction. However, it remains to be seen if cognitive and social responses, let alone a reduction in touch sensitivity, could be affected by this approach.

Like sections in an orchestra, our bodies are composed of instruments that take in a great deal of information in the form of different modalities and combinations of senses, including touch, smell, hearing, taste, and vision. Once an orchestra has tuned all of the instruments to a certain frequency, there is a more pleasing sound. Likewise, I believe our senses work optimally when they are at the right resonance.

So we have discussed how individuals diagnosed with autism experience the world according to their five (and six) senses. While each individual has a different experience, most children with autism have an extreme reaction to sensory stimulation, which affects each of the other senses. What if there were another (seventh!) sense that reports external and internal environmental information? Remember, the periphery is an integral part of the body's sensory experience, as is the immune system, which, as noted in Chapters 2 and 3, has been referred to as the "seventh sense."[45]

The Seventh Sense

As discussed in earlier chapters, the immune system is a network of cells and proteins that defends the body against all kinds of infections. We have previously outlined the periphery's role in an immune response due to its connections with the brain and the gut. Recall that in the case of an "invader," the peripheral nervous system senses bacteria through an armamentarium of T cells, B cells, macrophages, and dendritic cells. These dispatches—communication—find their way to the brain. The immune system also defends against invaders in the gut, where numerous bacteria and interactions can affect behavior. In a 2010 study of Danish children hospitalized for viral or bacterial infections, researchers found that a large number—more than 7,300—were diagnosed with autism, supporting a potential connection between infection and autism.[46]

Leo Kanner's first descriptions of autism in 1943 summarized not only sensory problems but also gastrointestinal issues, such as constipation, much like James Parkinson's description of shaking palsy. Recent studies have supported a correlation between autism disorders and the brain-gut axis. Over 50 percent of children diagnosed with autism are known to have problems associated with the gastrointestinal system, such as diarrhea and stomach pain.[47] It is also thought that some children may suffer from a leaky gut, a syndrome in which intestines become permeable and leak their contents into the bloodstream. We have already considered the idea that the microbiome, or its disruption, can contribute to the pathology of neurologic disorders like Parkinson's disease.[48] But the involvement of the gastrointestinal tract in autism is proven.[49] Although researchers continue to hypothesize whether gut bacteria initiate the disorder or are a consequence of the disorder, it is still a mystery.

In keeping with the current interest in the connection between gut microbes and behavior, researchers have begun to study events in an expecting mother's immune system that might lead to autism. One line of inquiry surrounding babies whose mothers had a viral infection during pregnancy showed that the production of cytokines and factors like TNF-α became elevated in offspring who were autistic. So a current theory is that children are more susceptible to autism from release of cytokines produced following an immune reaction.[50] Specifically, altered expression of immune reactive cytokines and markers of oxidative stress appear as a response to inflammation. Another explanation for the high incidence of autism points to illness in the mother. In a situation where there is maternal immune activation, an increase in the levels of inflammatory markers, such as in a viral infection during pregnancy, the high levels of stress and associated physiological changes can be passed on to the offspring. The rise in inflammatory cytokines can impact the vagal network (superhighway) of the fetus and can cause dysregulation of the brain's development; it can also affect the microbiota of the mother and the prenatal environment. Accordingly, the microbiome of the fetus can also be influenced.

Animal models of maternal infection also reveal that the offspring display characteristics of autism, associated with changes in the immune system. Sarkis Maxmanian and researchers at the California Institute of Technology induced autism-like symptoms in mice using an established paradigm of infecting mothers with a virus-like molecule during pregnancy. After birth, the mice had alterations in gut bacteria, but when treated with a health-promoting bacterium called *Bacteroides fragilis*, levels of anxiety and other behaviors improved.[51] Another scientist we encountered earlier, Paul Patterson, who proposed that glial cells were responsible for regeneration, developed several rodent models to study maternal infection as a risk factor for autism. His studies also mimicked infections by activating the maternal immune system without exposure to a pathogenic agent. Analysis of the offspring revealed considerable overlap in behavioral abnormalities and neuropathology similar to autistic traits. So the maternal influence connection in autism may not be as outlandish as it first seemed.

However, since our puzzle is not yet solved, especially as it relates to the periphery, I believe we need to consider a hormone associated with trust, happiness, and love. Oxytocin may initially seem like a far-fetched link to autism and other peripheral disorders, but the hormone brings a number of findings, theories, and pathways together.

Oxytocin

In 1906, Sir Henry Dale identified a small peptide hormone that caused the uterus to contract. He called it oxytocin, meaning "swift birth" in Greek. Years later, in 1953, Vincent du Vigneaud purified and biochemically synthesized it.[52] As a Nobel laureate in medicine working at Cornell Medical School, du Vigneaud found that oxytocin was produced in the hypothalamus and secreted by the pituitary gland, a pea-shaped structure at the base of the brain. A naturally produced hormone, oxytocin is most well known for managing labor, delivery, and lactation. During labor and childbirth, the hormone is released into the bloodstream in response to the uterine contractions and stretching of the cervix. It also stimulates suckling during breastfeeding. In fact, oxytocin is often called the "love" hormone, after its influence on mother-infant bonding.[53] Because it naturally induces labor and breastfeeding, the hormone has also been used medically to facilitate childbirth or to speed labor.

Investigators also found that other reactions, such as motivation, recognition of others, and trust, are also affected by this hormone. As in other scientific and medical studies, investigators sought not only human subjects but animals as well. In particular, voles, a relative of mice, have given researchers insight that is relevant to human behavior.

Prairie voles live in colonies in the Great Plains of the United States. Because their habitat had few food sources and a low density of animals, prairie voles evolved to ensure their species would live on by making bonds in pairs, monogamously. Besides forming lifelong bonds, each vole couple shared nesting and pup-raising responsibilities.[54] Yet other species of voles, such as the montane vole, are promiscuous and do not pair bond like the prairie vole. What accounts for this difference in social contact? The biological explanation for this behavior has been traced to the existence of oxytocin in female voles and, a close relative of oxytocin, the vasopressin hormone, expressed in males. Investigators found that beyond supporting social behaviors like ingroup bonding and maternal behavior, oxytocin is expressed in a number of organs, muscles, and tissues in the periphery, including the uterus, kidneys, pancreas, and thymus. Recent investigations have demonstrated that oxytocin receptors in the brain respond to the senses, often in line with the reward system.

We can see this connection with the sensory systems through reaction to stimuli. As an example, we could consider the bonding between pairs of mother and offspring satisfying the sense of touch. Oxytocin also has a role in memory, specifically, where to find food and whether a type of food is appealing, as taste buds can also be modulated by the hormone. It would also stand to reason that since the hormone increases bonding and trust, oxytocin would also facilitate an attentional bias toward familiar faces. Researchers have found that people induced with the hormone were more likely to look into the eyes of familiar faces, which emphasizes oxytocin's modulation of neural reward circuits and its role in trust and social support.[55] In humans, the left and right cerebral hemispheres of the brain are anatomically asymmetric and reflect functions of attention, memory, face recognition, and language. Interestingly, the receptors for oxytocin are arranged in the auditory system in an unexpected way: more oxytocin receptors are found on the left side of the auditory cortex than the right side.[56] Regardless of whether this placement helps to balance out our senses or enhances our ability to hear, a mother vole is more likely to retrieve newborn pups, since oxytocin assists her in hearing ultrasonic sounds. In sum, sensory cues—touch, taste, vision, and sound—are influenced by oxytocin.

I would be remiss not to mention that oxytocin can also alleviate pain.[57] Consider that the hormone is released during childbirth and that it induces labor, naturally preparing the body for delivery. By doing so, it also helps relieve the pain of childbirth. In fact, during most times of stress and discomfort, surges of oxytocin are released into the PNS and CNS through receptors located at multiple sites in the brain and throughout the spinal cord. In addition to decreasing pain signals and activating its receptors, the hormone binds to opioid receptors and stimulates opioid release in the brain.

Due to its profound effect on maternal care, bonding, and social attachment and enhancement of social interactions, as well as its ability to alleviate pain, oxytocin has been proposed as a treatment for anxiety and autism spectrum disorder.[58] In fact, clinical trials using a spray into the nasal cavity are currently ongoing.[59] Since sensory inputs are a source of problems—such as sensory overload, pain, and anxiety—for children with autism, it makes sense that this treatment might work.

The Peripheral Connection

So far, this book's journey has taken us through an exploration of several neurological diseases—Parkinson's, familial dysautonomia, Charcot-Marie-Tooth disorders—that share many symptoms arising from the peripheral nervous system, the immune system, and the gut. All the disorders have symptoms in common that affect sensory information and movement. But what about the additional difficulties of social interaction and communication and the repetitive behaviors for individuals with autism?

It is entirely plausible that defects in information processing in the periphery can explain or help to resolve these behaviors. After all, repetitive behaviors reflect a heightened or lowered threshold of the senses, and sensory symptoms are at the core of the development of social processing. For these reasons, one idea proposed is impaired sensory processing as a mechanistic link to autism.[60] After all, people navigate their environment with their hands, and physical contact activates oxytocin-producing neurons; even a light, gentle touch activates these neurons.[61] In fact, numerous studies have pointed to sensory processing as a core feature of autism.[62] So a straightforward strategy to address autism may be to combine touch, an increase of oxytocin, and enhanced social interaction as treatments.

Any plan should also consider our incredible human ability to change in changing circumstances. After all, the ability to adapt to new environments is a reason that humans and animals have been so successful and survived to the present. It just so happens that this ability to evolve and adapt is a bigger part of our bodies than most of us are aware, as we will explore in the next chapter on plasticity.

8

PLASTICITY IN THE PERIPHERY

> Plasticity, then, in the wide sense of the word, means the possession of a structure weak enough to yield to an influence, but strong enough not to yield all at once. Each relatively stable phase of equilibrium in such a structure is marked by what we may call a new set of habits. Organic matter, especially nervous tissue, seems endowed with a very extraordinary degree of plasticity of this sort.
>
> —WILLIAM JAMES

THE ABILITY TO CHANGE AND RESPOND to the environment, or to novel stimuli, is called plasticity. We often think about plasticity when considering the brain and recovery from injury, apropos learning, memory, and repairing or improving brain function. The subject of popular media articles and books, such as *The Brain that Changes Itself* and *Train Your Brain*, neuroplasticity and our ability to learn anew excite us.[1] But the brain is not the only part of our body capable of change. Plasticity also exists in the periphery.

In the 1880s, the philosopher William James (1842–1910) defied conventional thought and vocalized his theory that a neuron was constantly being remodeled throughout life. Living in New York City, William James was raised in an influential, highly educated family. His brother Henry became a famous novelist, but William was also an experimental writer, as his philosophical work presaged the current interest in the neural basis of creativity and experience. Trained originally at Harvard Medical School, William James is credited with starting the field of psychology, and he was also one of the first to think of behavioral outcomes of plasticity.

I am going to follow James's lead and propose that plasticity exists throughout the periphery. It is a controversial theory, an idea not promoted or studied in depth by the majority of neuroscientists, but I believe it is worth investigating. As we have seen throughout this book, the periphery is malleable and possesses the flexibility to respond to changes in the environment and to injury. The periphery, in the words of William James, is in "the possession of a structure weak enough to yield to an influence, but strong enough not to yield all at once."[2]

Modifications in the nervous system occur across a lifespan as a reaction to new experiences, unexpected events, and fluctuations in the environment. New pursuits or activity-dependent events—like learning a new language or traveling to a foreign country—enhance plasticity by forming new circuits, or "rewiring," and provide long-lasting effects associated with learning and memory.[3] In the CNS, neuronal activity is responsible for shaping the strength of synapses and the release of neurotransmitters. Signals from outside the cell can promote rapid responses in gene expression and protein synthesis, essential for growth and plasticity in the nervous system. Increases in synaptic activity raise the flux of calcium, which directs the expression of activity-dependent expression genes. As a result, there is a rewiring of neural circuits.

In a similar manner, the nervous system undergoes sensitive periods to accommodate and support renewal, like windows that open and close. During the "critical period," a restricted form of neuronal activity heightens intervals of malleability, which can affect language skills and the senses—hearing, touch, and vision. Outside these critical periods, the nervous system tends to become less pliable, to become, in effect, hardwired. In the field of linguistics, it is thought that language acquisition and vocabulary are obtained during the period from birth to age six.[4] The visual system also develops during this critical time. When the visual system forms, information from both eyes is integrated to give a single unified image. Acclaimed Nobel winners in medicine David Hubel and Torsten Wiesel found in their experiments with kittens that closing one eye in the first few months of life resulted in an irreversible loss in the closed eye; the deprived eye's connections to the brain shrank, and no amount of new experience could reverse the effects. Yet at the same time, the connections of the open eye became stronger. Remarkably, when the experiment was done in adult cats, vision was unaffected. Further experiments proved that disrupting eye development at a particularly young age had a detrimental effect on the visual circuit required for the maturation of vision. This was one of the first examples of how the critical period affects development.[5] We now know that plasticity during visual system development comes from the creation of new synapses and the growth of axons and dendrites. In the visual system, there is an absence of critical periods, so it becomes more stable and less plastic at later stages, similar to how most of the central nervous system works.

To understand the mechanisms of plasticity in the PNS, let's look closely at the effects of injury and changes in Schwann cells and see what kind of repair they are capable of.

Injury

Recall from the previous chapters that Schwann cells are a type of glial cell responsible for forming the myelin sheath around neurons in the PNS. While Schwann cells are generally not thought to transform like neurons, they still possess the property of plasticity. If there is an injury like a cut nerve, Schwann cells next to the injured nerve morph to become more like adult stem cells found in bone, muscle, or fat, supporting nerve repair by producing trophic factors and clearing myelin debris to foster a better environment for regeneration.[6] They proliferate and promote wound healing by facilitating repair and regeneration.[7] In other words, in response to injury, Schwann cells can switch to a persona that helps damaged neurons grow and supports their survival. These changes vary in time and number. Hundreds of Schwann cell genes can switch within a day of injury or over a period as long as several weeks, which suggests that a program of repair exists. But notably, as Schwann cells age, they have less plasticity to aid repair and regeneration; the ability to respond to injury and to clear debris is hampered. This explains why regeneration in the periphery decreases as the years add up; repair in the periphery diminishes with age.[8]

We previously discussed Schwann cells, which insulate the axons in nerve cells in the PNS, and the other type of glial cell, oligodendrocytes, which are responsible for the same role but in the CNS. But beyond the location, there is a stark difference in the ability to regenerate; as noted, Schwann cells evolve, reprograming to carry out new functions after peripheral nerves were cut. In contrast, oligodendrocytes are not capable of supporting regeneration in the CNS.

So why is the environment in the peripheral nervous system conducive to regeneration, particularly after injury, but not so for the central nervous system? One explanation for the CNS's inhibitory environment is that after development, the brain and spinal cord become hardwired and resistant to regeneration. But perhaps there are additional barriers, such as the formation of glia scars and increases in inhibitory proteins, that stand in the way of regeneration in the CNS. After all, these outcomes ensure the success of neurons and their targets; they are correctly assembled and fixed to give greater specificity. Further, during development there are safeguards to prevent immature neurons from becoming misplaced and misguided to the wrong environments (see Figure 6.1).

In the periphery, sensory nerves are pliable, so they are capable of regenerating. As an example, skin possesses many peripheral sensors and has a natural cycle of regeneration. Dead skin cells on the top layer of the epidermis fall away, and new cells underneath are ready to take their place.[9] We also see this kind of transformation in other senses as well, such as in structures like the taste buds and the olfactory apparatus, which both undergo a great deal of turnover during a lifetime. And the Merkel cells we encountered in the last chapter, which work closely with other sensory functions during touch, undergo continuous changes in shape and mechanical sensitivity throughout a lifetime, as do touch receptors, a proposed key factor in autism.[10]

Therefore, events in the periphery offer lessons to understand how the sensory system becomes receptive and can change its responses after changes in the environment. How can we harness this knowledge and promote plasticity elsewhere in the body?

Promoting Plasticity

Larry Katz (1956–2005), who served as the James B. Duke Professor of Neurobiology at Duke University, was a highly acclaimed neuroscientist supported by the Howard Hughes Medical Institute. Katz studied how changes in circuitry determined the development of the visual system, how the periphery responded to activity and sensory information, and how the effects of neural activity impacted the development of the nervous system.[11] Through his research, Dr. Katz realized that the nervous system was, in fact, highly capable of growing, adapting, and making new connections.

Shortly before he died, Katz, with colleague Manning Rubin, wrote a book for the general public called *Keep Your Brain Alive.*[12] Based on his investigations of growth factors in the periphery, Katz predicted that novel activities and exercises would aid in preventing degeneration of the nervous system. His recommendations were based on the abilities of neurotrophins—nerve growth factor (NGF) and brain-derived neurotrophic factor (BDNF)—to effect changes in membrane excitability and the configuration of neurons during development and adulthood.[13] Additionally, synapse size and numbers are affected, which led to the idea that trophic factors can translate activity into physiological forms of plasticity. Using this logic, Katz prescribed several simple exercises and tasks to enhance the nervous system, exercises with sensory stimulation and the kind of novelty that the brain craves during daily routines, such as taking a different route to work, brushing teeth with your nondominant hand, associating music with a smell, learning Braille or sign language, holding a blind wine tasting, showering with eyes closed, and reading aloud.[14] In this regard, travel, exercise, and new experiences have an effect on one's plasticity. Katz found that these kinds of new exercises and sensory associations also helped strengthen connections in the nervous system.

From Katz's work, we know that sensory stimulation leads to the production of neurotrophins (from the Greek word *trophe*, meaning to nourish), such as NGF and BDNF. Recall that NGFs are neurotrophic factors involved in the growth, maintenance, and survival of neurons. BDNF specifically induces and strengthens synapses and the growth of dendrites and assists in memory formation; it is considered a prime "plasticity" factor that is easily increased with new activity. According to Katz and Rubin:

> The amount of neurotrophins produced by nerve cells [and other cells like glia] . . . is regulated by how active those cells are. Meaning, the more active brain cells are, the more growth-stimulating molecules they produce and the better they respond. . . . Specific kinds of sensory simulation, especially nonroutine experiences that produce novel activity patterns in nerve cell circuits, can produce greater quantities of these growth-stimulating molecules.[15]

As stated above, Katz understood that a direct relationship existed between nerve cell activity and the production of growth factors.

In his research, he found that neurotrophins increased the size and complexity of dendrites that branch off neurons and that trophic factors like NGF could produce many more connections through new synapses.[16] Many later studies verified his theory that adding trophic factors to active neurons enhances dendritic growth. Additional confirmation of his theory comes when we look at the exact opposite scenario: When neurotrophins are removed, what happens? It turns out that nerve cells wither and die, leading to the decline often seen in neurological disorders.

Thanks to Katz's work, we know that the more active cells become, the more trophic molecules like NGF and BDNF are produced. Although these proteins were discovered over thirty years ago as brain hormones, we now know that they are produced throughout the body, even in peripheral organs. The principle is important, since trophic factors also play a role in preventing neurodegenerative diseases, such as Alzheimer's and Huntington's diseases, and in modulating pain and psychiatric disorders, such as depression and anxiety.[17] And BDNF and insulin-like growth factor-1 (IGF-1) have been found to improve memory and cognitive learning ability in the elderly.[18]

In 1982, Swiss neuroscientist Yves-Alain Barde was the first to identify, purify, and clone BDNF from a pig brain.[19] Given that BDNF is found only in meager amounts in the brain, this was a remarkable accomplishment. Barde's work paid off, as BDNF is a neurotrophic factor for many brain regions responsible for dopaminergic neurons, Alzheimer's disease, peptides responsible for respiration, glucose metabolism, and anxiety, and as a means for plasticity. BDNF also has interesting connections to obesity and diabetes. Investigators found that mutations in BDNF RNAs can produce aggressive behavior, depression, and eating disorders in mice and humans.[20] Further, mice and humans lacking BDNF become obese and are also susceptible to memory loss, in the case of Alzheimer's disease, as a result of the reduction in synaptic protein expression.[21] That is quite a response set off by a single protein!

Since the discovery of BDNF as a new neurotrophic factor in the NGF family, investigators have been attempting to uncover how any one single factor can induce plasticity, to improve learning and memory and prevent age-dependent cognitive decline. A primary focus has been on the signaling through the BDNF receptor, called TrkB, which activates several enzymatic pathways through modification of proteins by phosphorylation. The TrkB receptor, a crucial component in mood disorders and learning and memory, behaves as an antidepressant.[22]

Over the course of this book, I have used musical analogies to describe the organization of the peripheral nervous system. And researchers at MIT used a musical model to help explain how Schwann cells and oligodendrocytes differ in protein composition. Musical connections have been inferred with BDNF and its receptor through their increases after new activity. Brain levels could not be determined, but the inference is that music therapy can improve social interactions and quality of life. Incidental observations have shown that music is beneficial to the elderly and those suffering from dementia. Accordingly, the Orpheus Chamber Orchestra in New York has recently invited Alzheimer's patients to experience the healing power of music at some of their performances.[23]

But how do growth factors such as BDNF transmit their effects to the periphery? The answer is that different concentrations of BDNF, and even the timing of administration, give different signals and downstream signaling. One way is through endocrine effects, which generate neurotransmitters and small peptides made by the hypothalamus to regulate circuits and distant tissues dedicated to feeding. Lowering BDNF results in an increase in food intake and obesity, whereas raising BDNF levels results in reduced food intake. Thus, BDNF is capable of an endocrine effect on metabolism and growth in animals and humans.[24] The aftereffects of BDNF depend on its ability to trigger reactions that are transmitted to many peripheral tissues, such as muscle, liver, and fat.

Another way to exert an influence on the state of plasticity is through physical exercise. While we often consider exercise's beneficial impact on blood flow, heart rate, musculature, neurogenesis (production of new neurons from cell division), and resistance to injury, we may not realize that physical activity also produces the secretion of hormones, growth factors, and cytokines that affect nerve activity and brain function. These proteins are made in the periphery from many sources—skin, liver, muscle, cartilage—and assist in the survival of connections in the nervous system. Further, exercise is a potent natural inducer of growth factors that promote repair and regeneration, like the aforementioned BDNF and IGF-1, which is manufactured in the periphery in many different types of cells—endothelial cells, lymphocytes, platelets, megakaryocytes, and skeletal muscle. Many studies show that exercise is associated with synaptic plasticity and increases in neurogenesis.[25] In addition, burning sugars and fatty acids during physical exercise results in the production of small-molecule metabolites, such as lactate, which can enhance learning and memory, and by-products.[26] Lastly, all forms of exercise enhance the removal of toxic substances; physical exercise increases the flow of molecules in cerebrospinal glymphatic channels, which collect and recycle materials.[27] Exercise enhances the integrity of the cardiovascular system and removes the accumulation of toxic products under disease conditions. In short, exercise has enormous benefits for physical fitness and health, for an increased ability to heal, and for removal of toxins from the body.

Since novel experiences and exercise raise the level of trophic factors, which in turn prevents degeneration and helps neurons to be healthy, new forms of activity and physical exercise lower the risk for many disorders and diseases, including Parkinson's disease. Dance and music are particularly effective at improving balance and rhythmic motor coordination. The Mark Morris Dance Group in Brooklyn, New York, has found that dance does far more than enhance the physical body; it improves people's mental lives as well.

The Mark Morris Dance Group sponsors weekly dance classes and workshops specifically for Parkinson's patients.[28] Studies of Parkinson dance groups have found that the quality of life and motor control are enhanced after these classes. This is not surprising, since dance stimulates the body and the senses, simultaneously improving cognition, memory, and social interactions—promoting plasticity—with wide-ranging mental and physical health benefits.[29]

Other Stimulation

Thus far, the kinds of plasticity discussed have been those that are naturally occurring, relatively pleasant to experience, and generated from gentle means. There are more drastic ways to stimulate the nervous system. One is to apply a major shock to the system—to stimulate electrically, using electroconvulsive therapy. This kind of deep brain stimulation has been used in major depressive illnesses and Parkinson's disease.[30] It is plausible that enhanced neuronal activity stimulates the secretion of trophic factors or their receptors, so this kind of therapy could trigger TrkB signaling.[31]

There are ways to stimulate the spinal cord after a severe injury, to train the spinal cord to remember the pattern of walking. The spinal cord has an internal generator called a central pattern generator, which serves to remember the motion of walking.[32] When an injury severs the spinal cord, the central generator contains a memory of movement, and it can be revived by stimulation, reactivating nerves that have been asleep. To do so, an electrode is implanted or placed under the skin and operated using a remote control. As with deep brain stimulation, increasing the activity of the spinal cord enhances the ability to move the arms and legs after a serious injury. This kind of therapy is now being used for strokes and spinal cord injuries, a remarkable advance made possible by Albert Aguayo's original scientific findings showing that regeneration can be achieved after a severe injury to the spinal cord.

Although much of the last discussion focused on events that can occur in both the PNS and the CNS, that foundational information was another piece of the puzzle needed to fully realize the possibilities for plasticity in the periphery. Locomotor and treadmill training has emerged as an effective rehabilitation approach, where specific movements trigger sensory information that aids in walking and balance. Here is where we investigate the potential to improve health and to counter disease and aging.

Regeneration

I believe that the periphery holds the answers for regeneration and for reversing age-related declines in plasticity. And I am not alone in this thought. Because events in the periphery contribute strongly to the ability to recover from injury and neurodegeneration, new ideas about using peripheral cells to combat Alzheimer's and Parkinson's diseases, as well as aging, have proliferated.

One diverse approach to regenerative medicine is called parabiosis, from the Greek words *para* (adjacent to) and *bios* (life). The idea is simple: join old and young blood systems, and look for the peripheral factors that prevent age-related impairments. In other words, factors from young blood may help diseased or aged tissues to regenerate.[33] While that idea may sound like science fiction, experiments have shown that older brains can be rejuvenated by factors present in the blood of young animals and extend the life span of animals afflicted with disease. One experimental intervention involves surgically joining the circulatory systems of two animals, connecting blood vessels from old and young animals to identify molecules that are involved in the process of aging, looking for causes of age-related diseases.[34] Parabiosis is actually a tissue-grafting procedure. In 1864, the French physician Paul Bert wrote a thesis, *Experiences et considerations sur la greffe animale,* for a medical degree at the Faculté de Médecine de Paris.[35] As part of his dissertation experiment, two animals were joined through their skin and muscle walls. Bert showed that fluid injected into the vein of one animal could be simply passed to the other animal. Since the original experiment, the parabiosis procedure has been applied in many different ways, and with successful outcomes. In studies in which older animals were paired with young animals, the older animals lived longer. After young and old rats were connected by parabiosis, metabolic changes, like the turnover of cholesterol, were improved. In a recent study, aging contributed to a decline in the number of stem cells, but when joined to a younger environment, these stem cells regained a more youthful state, a result that suggested more mature cells or an older brain are capable of reversing age-related changes in a new and younger environment. And according to studies from several independent laboratories, aging becomes delayed in older mice exposed to a young mouse's circulatory system.[36] In a model of Alzheimer's transgenic mice, parabiosis reversed cognitive decline and improved working memory and cognitive functions.[37] Perhaps the most striking intervention, with respect to the brain, is intravenous administration of plasma, a soluble fraction of blood, from young mice, which was found to improve cognitive function in older mice. These surprising findings, like Aguayo's experiments with Schwann cells and the spinal cord discussed previously, indicate that the environment (and the local neighborhood) are critically important for transforming the aging process. However, not all tissues and organs were successfully rejuvenated through parabiosis, including the aging thymus. So it is not the method to fix it all.

A related, more widespread technique is the transfusion of blood. It is a safe clinical practice today and has been proposed for use in regenerative medicine, since young blood is a rich source of factors that could help revitalize the functions of an aging brain, muscles, and liver. Some researchers believe that there may be both positive and inhibitory factors that affect young versus old, although these factors have not all been identified. Nonetheless, several biotech startups have begun to test the ability of young blood to delay Alzheimer's disease.

The advantage of giving the blood of a young person to an old person is that it may contain rejuvenation factors that have multiple effects on many organs at the same time. Whole blood is a mixture of cells, chemicals, proteins, and breakdown products of metabolism (Figure 8.1), each of which can be separated from the others to identify their origins. These constituents represent a means to aid old tissues and to promote tissue regeneration and repair. Platelet-rich blood derivatives, such as platelet-rich plasma, produce many growth factors and hormones possessing survival and regenerative capacities that can help aging cells. For this reason, instead of blood transfusions, biotech companies like Alkahest are actively testing the potential of plasma proteins.

But before we get too excited about parabiosis with young mice or the transfer of young plasma, we must remember that the research is still in its early stages. There are likely many other age-related proteins with beneficial effects for the entire body that remain to be discovered. So far, we do not know how many factors are necessary to promote a youthful condition or how plasma from young animals is effective. And given the large number of immune and inflammatory responses during aging and neurodegeneration, it will be useful to identify more plasticity-enhancing factors and ways of promoting regeneration. The best scenario would be to use a patient's own plasma or platelet-derived cytokines and growth factors to stimulate healing and regeneration.

Trophic Factors

I believe an obvious approach is to increase the availability of trophic factors, such as BDNF, NGF, and IGF-1. It is conceivable that increasing availability of BDNF to counter aging may work, since as we know, BDNF is naturally highly expressed in the periphery, in

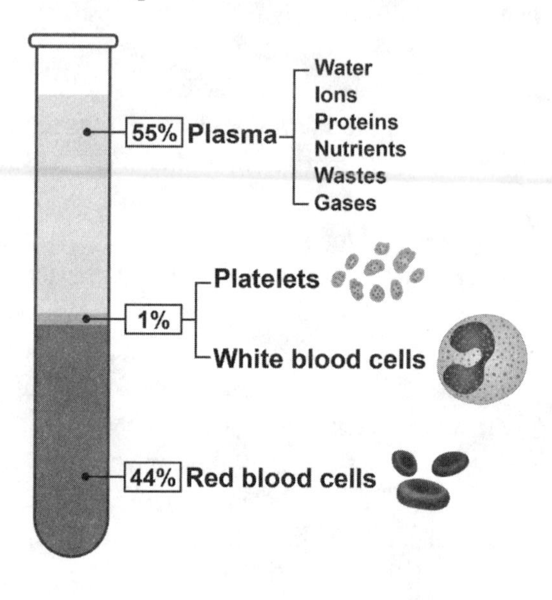

FIGURE 8.1 Blood consists of many ingredients—platelets, white blood cells, red blood cells, and plasma.

endothelial cells, astrocytes, lymphocytes, and muscle. It is also surprisingly abundant in blood platelets, making it easier to find and determine its concentration.[38] However, problems of administering BDNF to human subjects make it unlikely that BDNF will be directly introduced, and the difficulties of delivery, poor diffusion of neurotrophin proteins, and side effects in earlier neurology trials prevented further efforts using NGF and BDNF.[39] The pharmacokinetics of these proteins and the problems in managing the dose have hindered the application of neurotrophic factors as a therapeutic intervention in previous clinical trials.

Another surreptitious approach to the problem is to take advantage of small molecules made in the periphery. As mentioned in the previous discussion, metabolites are produced after exercise or fasting. While breakdown products usually become waste, these compounds do have a utility. Some of the molecules, called ketone bodies, made in huge quantities in the muscle and that live as by-products from exercise and fasting, migrate to many other organs in the body, including the brain.[40] They are small enough to pass through the blood-brain barrier and to provide an energy source, and they can significantly increase trophic factors like BDNF.[41] Perhaps platelet-derived neuronal differentiation in the periphery could serve as a vehicle for plastic changes in the brain.

We must also consider that trophic factors have been used to successfully treat Charcot-Marie-Tooth disease. Recall that CMT is a genetically inherited disorder that affects the Schwann cells, myelin sheath, axons, nerves, and muscles. As part of an experimental treatment, trophic factors were transferred into the muscle, which improved nerve conduction and increased myelination; the introduction of neurotrophins into muscle tissue had positive effects on myelin and axons from nearby nerves, as well as on Schwann cells.[42] So a single neurotrophic factor when placed in the right location can have dramatic consequences on a cell and many other cells surrounding it.

Another attractive approach is a combination of the previously noted ideas: to increase trophic factors from plasma. The factors identified in young plasma remind us that neurons do not function in isolation; they are part of a sophisticated network that includes glial cells, vascular cells, muscle, blood, and numerous other cell types. Increasing the trophic factors would allow the quick dispersion throughout the body. The techniques now exist to identify, test, and verify factors in circulation that can rejuvenate the nervous system during the aging process and after injury. We just need to experiment to see if they will work.

These are all exciting ideas and possibilities for how to magnify and expand the body's inherent capacity for regeneration. I have no doubt that in the coming years, we will see effective approaches, derived from these theories and experiments, to treat disorders and counter aging. Even complex diseases like Parkinson's, familial dysautonomia, CMT, and autism are ready for these kinds of targeted therapeutic efforts. Or perhaps a new idea that has not yet been considered will be the one to treat them all. Regardless, I believe the answers will come from the periphery.

CONCLUSION

> The central nervous system, along with the proprioceptive system, tells one who one is and what one is doing. . . . The autonomic nervous system evolved long before the central nervous system. . . . Sleeplessly monitoring every organ and tissue in the body, [it] tells one *how* one is. (Curiously, the brain itself has no sense organs, which is why one can have gross disorders here yet feel no malaise.)
>
> —OLIVER SACKS

IN ONE OF HIS LAST BOOKS, neurologist and author Oliver Sacks argued that the peripheral autonomic nerves have an important mission, independent from the brain. He perceived how important the periphery and sensory systems are, especially when it comes to awareness of the environment. Over the course of this book, I have argued the same.

In the field of neuroscience, the brain is considered the command post, the hub of power. Like the conductor who is united with all sections of the orchestra—strings, woodwinds, brass, and percussion—the brain is interconnected with the periphery and the rest of the body. The conductor responsible for leading the interpretation and pace of the music listens to the orchestra carefully. In a similar manner, the brain pays attention to the multitude of messages from many parts of the sensory system and guides the motor system.

The periphery is connected to this center through the five senses and many lines of communication. The PNS is adept at sensing both internal and external environments through the senses, including the proprioceptive nerves (although Sacks attached these nerves to the workings of the CNS, I have argued otherwise). With an ability to detect balance or a change in temperature by only a few degrees, to discriminate between different sugars and tastes, to hear varying pitches and rhythms, and to contrast unfamiliar smells, the peripheral nervous system is uniquely and highly responsive to its surroundings.

In this book, several unconventional ideas have been proposed, the first of which was that serious diseases of the nervous system start in the periphery. The traditional, mainstream view is that brain diseases are solely in the brain's domain, and the periphery has a role that is incidental at best. However, to form a complete understanding of the causes of and potential treatment for disease, we must look to the clues in the body.

The periphery's role in health and disease has been overlooked because its actions often work subconsciously. The periphery monitors the inner state of our organs and provides feedback without the need to be consciously aware. It is also governed by autonomous actions. When the PNS sends information between the body and the brain, we are usually not aware of the efforts, since they often occur without our need to guide them. When we pat a dog on the head, we do not need to tell the nerves to send information to the brain so we can make sense of the dog's soft fur. It is only when the peripheral nerves receive a warning (pain) or distort messages (resulting in muscle weakness) that we should pay attention. That is a signal to an individual that something is wrong; for a researcher or physician, it is an indication to pay attention and look more closely.

Although Parkinson's disease is classified as a brain disorder that progressively affects the nervous system, some harbingers of the disorder—constipation and difficulties with sleep—precede by at least ten years visible symptoms like tremors. And as we know, these portents all involve the PNS. Further, in contrast to the prevailing view that the pain center is solely in the brain, pain is also a peripheral problem, since it is unequivocally linked to the sensory feedback in the PNS.

Another example is the gut-brain connection—the gut microbiota associated with Parkinson's and other neuropsychiatric diseases. Now that the spread of disease proteins from the enteric nervous system to the CNS, once unheard of, is supported by a great deal of documentation, perhaps we will approach therapeutic strategies with the PNS in mind. It appears we may, as an evidence-based hunt for new targets and sites of action for delaying neurological diseases has accelerated to an allegro tempo.

Peripheral connections to numerous diseases—Parkinson's disease, familial dysautonomia, and autism—could be boiled down to the neurons or, to be more specific, a loss of neurons. So to counter and treat these disorders, we must figure out how to keep neurons alive. As we have seen, the periphery can regulate and regenerate; along with genetics and the environment, sensory neurons can increase or decrease lifespan. The senses detect incoming sensory signals and changes in the environment, so cues like pain promote neuroprotection, allowing animals and humans to survive stressful conditions. In essence, sensory neurons are responsible for survival and lifespan decisions. Therefore, the periphery should be the focus for lifespan and aging research.

Longevity has been a basic mystery in the life sciences. To discuss this potential, the following brings together all of our knowledge from the chapters—evolution, anatomical structure, role in disease, and the potential for regeneration and repair—to work on the problem of lifespan.

Lifespan

How lifespan is regulated is a unique and unsolved problem in biology. Our complex body system is composed of nerves, senses, organs, and muscles evolved from a worm, a simple organism that lacked a brain. The worm had only a "periphery," a nerve cord with groups of nerve cells that formed ganglia up and down the axis of the animal, and through these nerve nets, the animal was able to sense the environment, similar to our modern-day jellyfish. Then, as life became more complex, ganglia appeared in the front of the animal, a maneuver that resulted in the "brain." The organization of synapses and their circuit connections—what neurons need to communicate with each other and other cells—formed as part of the network. It is this organization that occurred back in the early stages of evolution, rather than size, that dictates higher-order behaviors.

It is appropriate, then, to close the discussion of the length of life through investigation using a worm. Researchers used a simple model organism called *C. elegans*, a nematode or round worm, to study lifespan because these worms are short-lived, and it is easy to distinguish the phases of life, physiologically and genetically. As one would expect, the animals are equipped with well-defined sensory networks that help them properly detect and respond to environmental factors. We have learned that of the 302 nerve cells in *C. elegans*, sixty are classified as sensory in function. Investigations revealed that lifespan can shift over ten times during the course of their short life due to their genetic makeup and in response to fluxes in dietary restriction and nutrients, temperature, and even signals from the reproductive system.[1]

Investigators identified specific genes involved in longevity, many of which are located in sensory neurons we discussed earlier. The first pathway associated with regulation of lifespan is an evolutionarily conserved receptor and its ligand, the insulin-like growth factor IGF-1.[2] Recall that in the previous chapter during our plasticity discussion, we encountered insulin and IGF-1, an insulin family member and hormone that promotes the growth of bones and tissues. Research in the fruit fly, *Drosophila melanogaster*, also demonstrated that insulin/insulin-like peptides exist in parallel pathways that participate in the determination of lifespan.[3] For humans, symptoms of excess in this hormone can lead to obesity or excessive growth of a certain part of the body, such as the hands or head. But mutations in this growth-factor system can increase lifespan in humans, although doing so requires participation from a transcription factor called DAF-16 under the direct control of insulin/IGF-1.[4] Without going too far into these particular factors, which would take us away from our main discussion, the important result in this case is that the pattern of gene expression is changed in the pathway.

As noted in the lifespan findings of *C. elegans*, dietary restriction and nutrients are also linked to longevity. The selective advantage of the insulin/IGF-1 system is, in fact, the ability to detect nutrients. This is also the case in the human body, which is where the excess supply of the hormone can lead to obesity. The beneficial effects of food level detection and fasting or calorie restriction have also been traced to sensory pathways that use the IGF-1 pathway.[5] The sensory influence on lifespan is depicted in Figure C.1.

Let's consider this process in the model organism *C. elegans* out in the wild—in a temperate soil environment—rather than in the lab. When something occurs in that environment, say a piece of rotting fruit being added to a compost pile, the organism uses sensory neurons to detect and process the environmental cues, in this case, olfactory perception of food. The olfactory neurons perceive food odor, transmit the cues, and assess food quality, then regulate the feeding behavior, resulting in increased health and lifespan.

Many observations point to a strong sensory influence on longevity. The sensory effects on lifespan start with processing environmental cues, which are transmitted to neural circuits and tissues that dictate survival and longevity. As sensory neurons are located and associated in the periphery, the significance of the peripheral nervous system is underscored.

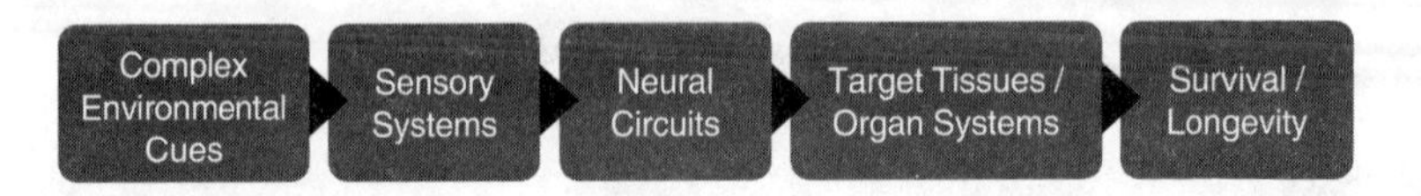

FIGURE C.1 Flowchart depicting how the environment and sensory networks fit into the longevity problem.

Temperature (or heat)—external and internal—also appears to be a potent signal to modulate longevity. As stated previously, researchers found that temperature is a factor in the lifespan of *C. elegans*. Studies have also shown that thermosensory neurons, which process genetic and environmental cues, affect the lifespan of *Drosophila melanogaster*, the fruit fly, as a model organism. More broadly, many mutations that extend lifespan affect stress-response genes or nutrient sensors; dietary restriction is a well-known environmental signal that increases lifespan in many species, and chemosensory signals from the reproductive system and reductions in the rates of respiration or translation influence aging in some species.

Here we must include the capsaicin receptor TRPV1. You may recall that this protein is responsible for detecting and regulating body temperature (nociception). In a recent study, genetically engineered (knockout) mice that lack TRPV1 displayed a pronounced lengthening of lifespan, well over one hundred days, which is a long extension for mice.[6] This was accompanied by observations of reduced aging, including youthful metabolism and an improved response to insulin. How can we explain these results? TRPV1 is an important indicator of hypersensitivity to heat, inflammation, and pain, and it is sensitive to many different proteins, lipids, and chemicals. For this reason, one explanation for the findings—the link between absence of TRPV1 and increased longevity—is that reduced pain detection has a direct effect on increasing lifespan.[7] This makes some sense, as pain can cause stressful conditions. It is postulated that TRPV1 ion channels in sensory neurons may be a protective mechanism that serves as a signal for harmful effects. Absence of TRPV1 is postulated to promote longevity, perhaps through changes in metabolic activity.

The investigators took it one step further. TRPV1 mutant mice—that is, mice with an absence of TRPV1—have an increased lifespan, a youthful metabolism, and an improved spatial memory, as well as insulin resistance.[8] These findings are more difficult to explain, since TRPV1-positive sensory neurons are designed to detect life-threatening signals and to coordinate different neuronal signals in the periphery. But it is clear that detection of environmental cues like temperature, whether from a sensory neuron or ion channel, were needed for evolutionary survival in stressful environments.

To consider how these neurons and ion channels' signals develop in a disease state, we will consider familial dysautonomia, the genetic disorder that affects the sensory and autonomic systems. In Chapter 6 we discussed how individuals with familial dysautonomia are sensitive to pain and temperature. Important to the longevity discussion, this disorder represents one of the most serious forms of hereditary peripheral neuropathies; only 60 percent of patients survive to twenty years of age.

The elongator protein ELP1 was identified as the gene mutation that causes familial dysautonomia; a person with the disorder does not produce enough of the protein. Researchers observed that mutations in ELP1 cause the death of sensory and sympathetic neurons, which then has a broad effect on many cellular events, such as chromatin structure and the activity of RNA polymerase, two essential functions involved in generating proteins.[9] This cascading effect can also result in systemic problems in the patient's cardiovascular system, kidneys, GI tract, and lungs. Even control of involuntary actions such as breathing, blood pressure, digestion, and temperature can be affected.

As mentioned previously, nerve growth factor (NGF) signaling has been closely linked to ELP1, and in ELP1 mutations there is a decrease in proper transport of NGF from the affected sympathetic neurons. Therefore, it is conceivable that a lack of trophic factors such as NGF is associated with the loss in longevity through mutations in ELP1 and perhaps affects lifespan more broadly. After all, the mechanism of pain initiated in the periphery also involves NGF in sympathetic neurons, and it is logical that a lack of growth factor means death.

Here we are at the end of our consideration of lifespan, but back to the initial statement of how the periphery connects to numerous diseases can be simplified to a loss of sensory neurons. Although we do not know all the mysteries of lifespan, we do know that during development, at the beginning of the life cycle, nearly 50 percent of sympathetic neurons are lost during a phase called naturally occurring, or programmed, cell death. In this way, naturally occurring cell death shapes the formation and longevity of the nervous system, but greater numbers of sensory and sympathetic neurons will die if NGF or other trophic factors are removed; in the absence of trophic support, rapid cell death occurs. It really comes down to the initial phase when a neuron becomes attached to a single trophic factor.

Despite years of research, there is a gap in our knowledge about exactly how neurons become attached to a necessary growth factor. Little is known about how neurons become dependent on or independent of trophic support such as from NGF or IGF-1. We just know that after removal of trophic factors, neurons rapidly undergo programmed cell death, or apoptosis. And how neurons avoid cell death after a loss of trophic support has yet to be determined, although it is likely to be related to longevity. Several examples of trophic factor independence exist in the nervous system, suggesting that maturity of neurons has an influence on the ability to fend off cell death.[10]

In a miraculous finding, sympathetic neurons maintained in culture for long periods of time become resistant to cell death, but the basis of this switch in dependence on trophic factors is not known.[11] On the other hand, neurons that become independent of trophic factor may contribute to longevity. Solving this riddle will provide much-needed insight into how aging neurons avoid cell death when there is a loss of trophic factors. We must seek the reason for mature neurons' increased resistance to trophic factor deprivation, as it is relevant to lifespan and to many of the neurodegenerative diseases that we discussed in this book. The field of genetics was unified only when the concept of the structure of the DNA double helix was deduced by Watson and Crick in 1953, a finding that revolutionized the study of biology and human diseases from cancer to virology. The field of neurobiology awaits such a unifying theory to connect its fundamental principles.

But we have discussed numerous guiding principles that are fundamental to the periphery and that address fundamental questions about the nervous system and disease. To underemphasize the periphery is to ignore the future of medical and scientific research and forgo or delay the probable insight that will lead us to a comprehensive understanding.

An orchestra without a string section is no longer an orchestra. It is still a beautiful ensemble, but it misses the sweeping lyrical line of the strings that connects all the sections. Similarly, without the periphery, we would totally miss the relay of communication from the brain to the spinal cord to the far reaches of the body. It is time for the idea of the peripheral nervous system as marginal or secondary to be cast aside. It is time we recognized the periphery as a vital and central part of the body whose impact extends even to longevity.

What is the big picture that emerges from discussing the periphery and its relationship to longevity? We have argued that lifespan is highly dependent on sensory and sympathetic neuron function.

Aging is a universal biological process that involves many factors that are inextricably linked to neurodegenerative disorders. Despite research mentioned here, there is still a gap in our knowledge about how neurons are able to retain longevity. Within the bounds of possibility, a case can be made that aging and longevity are influenced by the periphery.

Notes

Acknowledgments

Credits

Index

NOTES

Introduction

Epigraph: Irving Langmuir, Nobel Banquet speech, December 10, 1932, NobelPrize.org, www.nobelprize.org/prizes/chemistry/1932/langmuir/speech/.

1. Jon Palfreman, *The Race to Unlock the Mysteries of Parkinson's Disease* (New York: Farrar, Stauss and Giroux, 2015).

2. L. Norcliffe-Kaufmann, F. B. Axelrod, and H. Kaufmann, "Developmental Abnormalities, Blood Pressure Variability and Renal Disease in Riley-Day Syndrome," *Journal of Human Hypertension* 27, no. 1: 51–55, doi: 10.1038/jhh.2011.107.

1. The First System

Epigraph: Lewis Thomas, *The Lives of a Cell: Notes of a Biology Watcher* (New York: Penguin Random House, 1974), 3.

1. Amy Maxmen, "Marine Worm Rewrites Theory of Brain Evolution," *Nature* (2012): 10226, doi: 10.1038/nature.2012.10226.

2. Hiroaki Nakano, Kennet Lundin, Sarah J. Bourlat, Maximilian J. Telford, Peter Funch, Jens R. Nyengaard, Matthias Obst, and Michael C. Thorndyke, "Xenoturbella bocki Exhibits Direct Development with Similarities to Acoelomorpha," *Nature Communications* 4 (2013): 1537, doi: 10.1038/ncomms2556.

3. John N. Wood, "Nerve Growth Factor and Pain," *New England Journal of Medicine* 363, no. 16 (2010): 1572–73, doi: 10.1056/NEJMe1004416.

4. Patrick W. Mantyh, Martin Koltzenburg, Lorne M. Mendell, Leslie Tive, and David L. Shelton, "Antagonism of Nerve Growth Factor-TrkA Signaling and the Relief of Pain," *Anesthesiology* 115, no. 1 (2011): 189–204, doi: 10.1097/ALN.0b013e31821b1ac5.

5. Michael D. Gershon, *The Second Brain: A Groundbreaking New Understanding of Nervous Disorders of the Stomach and Intestine* (New York: HarperCollins, 1998).

6. Junjie Qin, Ruiqiang Li, Jeroen Raes, Manimozhiyan Arumugam, Kristoffer Solvsten Burgdorf, Chaysavanh Manichanh, Trine Nielsen, et al., "A Human Gut Microbial Gene Catalogue Established by Metagenomic Sequencing," *Nature* 464, no. 7285 (2010): 59–65, doi: 10.1038/nature08821.

7. Gil Sharon, Timothy R. Sampson, Daniel H. Geschwind, and Sarkis K. Mazmanian, "The Central Nervous System and the Gut Microbiome," *Cell* 167 (2016): 915–32, doi: 10.1016/j.cell.2016.10.027.

8. Meenakshi Rao and Michael D. Gershon, "The Bowel and Beyond: The Enteric Nervous System in Neurological Disorders," *Nature Reviews Gastroenterology and Hepatology* 13, no. 9 (2016): 517–28, https://doi.org/10.1038/nrgastro.2016.107.

9. Grant Anderson, Ali Reza Noorian, Georgia Taylor, Mallappa Anitha, Doug Bernhard, Shanthi Srinivasan, and James G. Greene, "Loss of Enteric Dopaminergic Neurons and Associated Changes in Colon Motility in an MPTP Mouse Model of Parkinson's Disease," *Experimental Neurology* 207, no. 1 (2007): 4–12, https://doi.org/10.1016/j.expneurol.2007.05.010.

10. Bryan B. Yoo and Sarkis K. Mazmanian, "The Enteric Network: Interactions between the Immune and Nervous Systems of the Gut," *Immunity* 46, no. 6 (2007): 910–26, doi: 10.1016/j.immuni.2017.05.011.

11. Ignacio Renbollo, Anne-Dominique Devauchelle, Benoît Béranger, and Catherine Tallon-Baudry, "Stomach-Brain Synchrony Reveals a Novel, Delayed-Connectivity Resting-State Network in Humans," *eLife* 7 (March 2018), https://doi.org/10.7554/eLife.33321.

12. Gershon, *The Second Brain.*

2. A Gut Reaction

Epigraph: Giulia Enders, *Gut: The Inside Story of Our Body's Most Underrated Organ* (Berkeley, CA: Greystone, 2018), 18.

1. Rodger A. Liddle, "Parkinson's Disease from the Gut," *Brain Research* 1693 (2018): 201–6, doi: 10.1016/j.brainres.2018.01.010.

2. Yuuki Obata and Vassilis Pachnis, "The Effect of Microbiota and the Immune System on the Development and Organization of the Enteric Nervous System," *Gastroenterology* 151, no. 5 (2016): 836–44, doi: 10.1053/j.gastro.2016.07.044.

3. Tomasz Brudek, "Inflammatory Bowel Diseases and Parkinson's Disease," *Journal of Parkinson's Disease* 9 (2019): S331–44, doi: 10.3233/JPD-191729.

4. Dan R. Littman and Eric G. Pamer, "Role of Commensal Microbiota in Normal and Pathogenic Host Immune Responses," *Cell Host Microbe* 10, no. 4 (2011): 11–323, doi: 10.1016/j.chom.2011.10.004.

5. David Brooks, "The Wisdom Your Body Knows," *New York Times* (November 26, 2019).

6. Staffan Holmqvist, Oldriska Chutna, Luc Bousset, Patrick Aldrin-Kirk, Wen Li, Tomas Björklund, Zhan-You Wang, Laurent Roybon, Ronald Melki, and Jia-Yi Li, "Direct Evidence of Parkinson Pathology Spread from the Gastrointestinal Tract to the Brain in Rats," *Acta Neuropathology* 128, no. 6 (2014): 805–20, doi: 10.1007/s00401-014-1343-6.

7. Norihito Uemura, Maiko T. Uemura, Kelvin C. Luk, Virginia M-Y Lee, and John Q. Trojanowski, "Cell-to-Cell Transmission of Tau and α-Synuclein," *Trends in Molecular Medicine* 16 (2020): 936–52, doi: 10.1016/j.molmed.2020.03.012.

8. Lisa Klingelhoefer and Heinz Reichmann, "Pathogenesis of Parkinson's Disease—The Gut-Brain Axis and Environmental Factors," *Nature Reviews Neurology* 11, no. 11 (2015): 625–36, doi: 10.1038/nrneurol.2015.197 .

9. Stanley B. Prusiner, "Biology and Genetics of Prions Causing Neurodegeneration," *Annual Review of Genetics* 47 (2013): 601–23, doi: 10.1146/annurev-genet-110711-155524.

10. Pawel P. Liberski, Agata Gajos, Beata Sikorska, and Shirley Lindenbaum, "Kuru, the First Human Prion Disease," *Viruses* 11, no. 3 (2019): doi: 10.3390/v11030232.

11. Simon Makin, "Pathology: The Prion Principle," *Nature* 538, no. 7626 (2016): S13–16, doi: 10.1038/538S13a.

12. Kevin C. Luk, Victoria Kehm, Jenna Carroll, Bin Zhang, Patrick O'Brien, John Q. Trojanowski, and Virginia M-Y Lee, "Pathological α-Synuclein Transmission Initiates Parkinson-Like Neurodegeneration in Nontransgenic Mice," *Science* 338, no. 6109 (2012): 949–53, doi: 10.1126/science.1227157.

13. Flavio Amaral, Daniel Sachs, V. V. Costa, Caio T. Fagundes, Daniel Cisalpino, Thiago M. Cunha, Sergio Henrique Ferreira, et al., "Commensal Microbiota Is Fundamental for the Development of Inflammatory Pain," *Proceedings of the National Academy of Sciences* 105, no. 6 (2008): 2193–97, doi: 10.1073/pnas.0711891105.

3. Pain Perception

Epigraph: Aristotle, *Aristotle's Art of Rhetoric*, translated by Robert C. Bartlett (Chicago: University of Chicago Press, 2019), 54b11.

1. Sigmund Freud, *Beyond the Pleasure Principle* (New York: W. W. Norton, 1955).

2. Allan I. Basbaum, Diane M. Bautista, Grégory Scherrer, and David Julius, "Cellular and Molecular Mechanisms of Pain," *Cell* 139, no. 2 (2009): 267–84, doi: 10.1016/j.cell.2009.09.028.

3. Diane M. Bautista, "Spicy Science: David Julius and the Discovery of Temperature-Sensitive TRP Channels," *Temperature* 2, no. 2 (2015): 135–41, doi: 10.1080/23328940.2015.1047077.

4. Natalie Richards and S. B. McMahon, "Targeting Novel Peripheral Mediators for the Treatment of Chronic Pain," *British Journal of Anaesthesia* 111, no. 1 (2013): 46–51, doi: 10.1093/bja/aet216.

5. Michael Costigan, Joachim Scholz, and Clifford J. Woolf, "Neuropathic Pain: A Maladaptive Response of the Nervous System to Damage," *Annual Reviews Neuroscience* 32 (2009): 1–32, doi: 10.1146/annurev.neuro.051508.135531/.

6. S. M. Collins, "The Intestinal Microbiota in the Irritable Bowel Syndrome," *International Review of Neurobiology* 131 (2016): 247–61, doi:10.1016/bs.irn.2016.08.003.

7. Bautista, "Spicy Science."

8. Michael J. Caterina, Mark A. Schumacher, Makoto Tominaga, Tobias A. Rosen, Jon D. Levine, and David Julius, "The Capsaicin Receptor: A Heat-Activated Ion Channel in the Pain Pathway," *Nature* 289, no. 6653 (1997): 816–24, doi: 10.1038/39807.

9. Ibid.

10. Jiale Luo, Jing Feng, Shenbin Liu, Edgar T. Walters, and Hongzhen Hu, "Molecular and Cellular Mechanisms that Initiate Pain and Itch," *Cellular and Molecular Life Sciences* 72, no. 17 (2015): 3201–23, doi: 10.1007/s00018-015-1904-4.

11. John N. Wood, "Nerve Growth Factor and Pain," *New England Journal of Medicine* 363 (2010): 1572–73.

12. Franziska Denk, David L. Bennett, and Stephen B. McMahon, "Nerve Growth Factor and Pain Mechanisms," *Annual Review of Neuroscience* 40 (2017): 307–25, doi: 10.1146/annurev-neuro-072116-031121.

13. Moses V. Chao, "A Conversation with Rita Levi-Montalcini," *Annual Review of Physiology* 72 (2010): 1–12, doi: 10.1146/annurev-physiol-021909-135857.

14. Denk, Bennett, and McMahon, "Nerve Growth Factor."

15. Yuki Terada, Shoko Morita-Takemura, Ayami Isonishi, Tatsuhide Tanaka, Hiroshi Okuda, Kouko Tatsumi, Takeaki Shinjo, Masahiko Kawaguchi, and Akio Wanaka, "NGF and BDNF Expression in Mouse DRG after Spared Nerve Injury," *Neuroscience Letters* 686 (2018): 67–73, doi: 10.1016/j.neulet.2018.08.051.

16. Rachael M. Easton, Thomas L. Deckwerth, Alexander Sh. Parsadanian, and Eugene M. Johnson Jr., "Analysis of the Mechanism of Loss of Trophic Factor Dependence Associated with Neuronal Maturation: A Phenotype Indistinguishable from Bax Deletion," *Journal of Neuroscience* 17, no. 24 (1997): 9656–66, doi: 10.1523/JNEUROSCI.17-24-09656.1997.

17. Nancy E. Lane, Thomas J. Schnitzer, Charles A. Birbara, Masoud Mokhtarani, David L. Shelton, Mike D. Smith, and Mark T. Brown, "Tanezumab for the Treatment of Pain from Osteoarthritis of the Knee," *New England Journal of Medicine* 363, no. 16 (2010): 1521–31, doi: 10.1056/NEJMoa0901510.

18. Wood, " Nerve Growth Factor and Pain."

19. Patrick W. Mantyh, Martin Koltzenburg, Lorne M. Mendell, Leslie Tive, and David L. Shelton, "Antagonism of Nerve Growth Factor-TrkA Signaling and the Relief of Pain," *Anesthesiology* 115, no. 1 (2011): 189–204, doi: 10.1097/ALN.0b013e31821b1ac5.

20. Huai-hu Chuang, Elizabeth D. Prescott, Haeyoung Kong, Shannon Shields, Sven-Eric Jordt, Allan I. Basbaum, Moses V. Chao, and David Julius, "Bradykinin and NGF Release the Capsaicin Receptor from Ptdins(4,5)P2-mediated Inhibition," *Nature* 411, no. 6840 (2001): 957–62, doi: 10.1038/35082088; Jan Minde, Göran Toolanen, Thomas Andersson, Inger Nennesmo, Aim Nilsson Remahl, and Olle Svensson, "Familial Insensitivity to Pain (HSAN V) and a Mutation in the *NGFB* Gene. A Neurophysiological and Pathological Study," *Muscle and Nerve* 30, no. 6 (2004): 752–60, doi: 10.1002/mus.20172.

21. Simona Capsoni, Sonia Covaceuszach, Sara Marinelli, Marcello Ceci, Antonietta Bernardo, Luisa Minghetti, Gabriele Ugolini, Flaminia Pavone, and Antonino Cattaneo, "Taking Pain out of NGF: A 'Painless' NGF Mutant, Linked to Hereditary Sensory Autonomic Neuropathy Type V, with Full Neurotrophic Activity," *PLoS One* 6 (2011): e17321, https://doi.org/10.1371/journal.pone.0017321.

22. Yashido Indo, "Neurobiology of Pain, Interoception and Emotional Response: Lessons from Nerve Growth Factor-Dependent Neurons," *European Journal of Neuroscience* 39, no. 3 (February 4, 2014): 375–91, doi: 10.1111/ejn.12448.

4. A Proneness to Trembling

Epigraph: James Parkinson, *An Essay on the Shaking Palsy* (London: Sherwood, Neely, and Jones, 1817); reprinted in *Journal of Neuropsychiatry and Clinical Neuroscience* 14, no. 2 (Spring 2002): 225.

1. Judith Flanders, *The Victorian City, Everyday Life in Dickens' London* (New York: St. Martin's Griffin, 2015).

2. André Parent, "A Tribute to James Parkinson," *Canadian Journal of Neurological Science* 45, no. 1 (2018): 83–89, doi: 10.1017/cjn.2017.270.

3. Patrick A. Lewis, "James Parkinson: The Man Behind the Shaking Palsy," *Journal of Parkinson's Disease* 2, no. 3 (2012): 181–87, doi: 10.3233/JPD-2012-012108.

4. E. D. Louis, "The Shaking Palsy, the First Forty-Five Years: A Journal through the British Literature," *Movement Disorders* 12, no. 6 (1997): 1068–72, doi: 10.1002/mds.870120638; Jon Palfreman, *Brain Storms: The Race of Unlock the Mysteries of Parkinson's Disease* (New York: Farrar, Strauss and Giroux, 2015).

5. Frank R. Freemon, "Galen's Ideas on Neurological Function," *Journal of the History of Neuroscience* 3, no. 4 (1994): 263–71, doi: 10.1080/09647049409525619.

6. Walter Isaacson, *Leonardo da Vinci* (New York: Simon and Schuster, 2017).

7. Gerald Stern, "Did Parkinsonism Occur before 1817?," *Journal of Neurology, Neurosurgery, and Psychiatry*, special supplement 11–12 (1989), doi: 10.1136/jnnp.52.suppl.11.

8. Ragnar Stien, "Shakespeare on Parkinsonism," *Movement Disorders* 20, no. 6 (2005): 768–71, doi: 10.1002/mds.20434.

9. Lance Fogan, "The Neurology in Shakespeare," *Archives of Neurology* 46 (1989): 922–24, doi: 10.1001/archneur.1989.00520440118029; Christopher G. Goetz, "Shakespeare in Charcot's Neurologic Teaching," *Archives of Neurology* 45, no. 8 (1988): 920–21, doi: 10.1001/archneur.1988.00520320122028.

10. William Shakespeare, *King Henry IV*, 2.4.7.89–90. References are to part, act, scene, and line.

11. William Shakespeare, *Troilus and Cressida*, 1.2.172–75.

12. William Shakespeare, *Richard II*, 2.3.103–3.

13. Patrick Haggard and Sam Rodgers, "Movement Disorder of Nicolas Poussin (1594–1665)," *Movement Disorders* 15, no. 2 (2000): 328–34, doi: 10.1002/1531-8257(200003)15:2<328::aid-mds1021>3.0.co;2-y.

14. Parkinson, "An Essay on the Shaking Palsy."

15. R. Horowski, L. S. Horowski, S. Vogel, W. Poewe, and F. W. Kielhorn, "An Essay on Wilhelm von Humboldt and the Shaking Palsy: First Comprehensive Description of Parkinson's Disease by a Patient," *Neurology* 45 (1995): 565–68.

16. S. Y. Tan and D. Shigaki, "Jean Martin Charcot (1825–1893): Pathologist Who Shaped Modern Neurology," *Singapore Medical Journal* 48, no. 5 (2007): 383–84.

17. Ibid.; Venita Jay, "The Legacy of Jean Martin Charcot," *Archives of Pathology and Laboratory Medicine* 124, no. 1 (2000): 10–11, doi: 10.5858/2000-124-0010-TLOJMC.

18. Christopher G. Goetz, "Amyotrophic Lateral Sclerosis: Early Contributions of Jean-Martin Charcot," *Muscle & Nerve* 23, no. 3 (2000): 336–43, doi: 10.1002/(SICI)1097-4598(200003)23:3<336::AID-MUS4>3.0.CO;2-L.

19. David R. Kumar, Florence Aslinia, Steven H. Yale, and Joseph J. Mazza, "Jean-Martin Charcot: The Father of Neurology," *Clinical Medicine and Research* 9, no. 1 (2011): 46–49, doi: 10.3121/cmr.2009.883.

20. Christopher G. Goetz, "Charcot on Parkinson's Disease," *Movement Disorders* 1, no. 1 (1986): 27–32, doi: 10.1002/mds.870010104.

21. Ibid.

22. Amit Zeisel, Hannah Hochgerner, Peter Lönnerberg, Anna Johnsson, Fatima Memic, Job van der Zwan, Martin Häring, et al., "Molecular Architecture of the Mouse Nervous System," *Cell* 174, no. 4 (2018): 999–1014, doi: 10.1016/j.cell.2018.06.021.

23. H. Braak, U. Rub, W. P. Gai, and K. Del Tredici, "Idiopathic Parkinson's Disease: Possible Routes by which Vulnerable Neuronal Types May Be Subject to Neuroinvasion by an Unknown Pathogen," *Journal of Neural Transmission* 110, no. 5 (2003): 517–36, doi: 10.1007/s00702-002-0808-2.

24. Ibid.

25. Elizabeth Svennson, Erzébet Horváth-Puhó, Reimer W. Thomsen, Jens Christian Djurhuus, Lars Pedersen, Per Borghammer, and Henrik Toft Sørensen, "Vagotomy and Subsequent Risk of Parkinson's Disease," *Annals of Neurology* 78, no. 4 (2015): 522–29, doi: 10.1002/ana.24448.

26. Timothy R. Sampson, Justine W. Debelius, Taren Thron, Stefan Janssen, Gauri G. Shasti, Zehru Esra Ilhan, et al., "Gut Microbiota Regulate Motor Deficits and Neuroinflammation in a Model of Parkinson's Disease," *Cell* 167, no. 6 (2016): 1469–80, doi: 10.1016/j.cell.2016.11.018.

27. Bryan A. Killinger, Zachary Madaj , Jacek W. Sikora, Nolwen Rey, Alec J. Haas, Yamini Vepa, Daniel Lindqvist, et al., "The Vermiform Appendix Impacts the Risk of Developing Parkinson's Disease," *Science Translational Medicine* 10, no. 465 (2018): eaar S280, doi: 10.1126/scitranslmed.aar5280.

28. Ibid.

29. Ibid.

30. Sampson et al., "Gut Microbiota."

31. Gil Sharon, Timothy R. Sampson, Daniel H. Geschwind, and Sarkis K. Mazmanian, "The Central Nervous System and the Gut Microbiome," *Cell* 167, no. 4 (2016): 915–32, doi: 10.1016/j.cell.2016.10.027.

32. Parkinson, "An Essay on the Shaking Palsy."

33. Frank Dikötter, *Mao's Great Famine: The History of China's Most Devastating Catastrophe, 1958–1962* (New York: Walker, 2010).

34. Li Zhisui, *The Private Life of Chairman Mao: The Memoirs of Mao's Personal Physician Dr. Li Zhisui* (New York: Random House, 1994).

35. Jung Chang and Jon Halliday, *Mao: The Unknown Story* (New York: Anchor Books, 2005).

36. James A. Horne, "Human REM Sleep: Influence on Feeding Behavior, with Clinical Implications," *Sleep Medicine* 16, no. 8 (2015): 910–16, doi: 10.1016/j.sleep.2015.04.002.

37. Marie L. Jacobs, Yves Dauvilliers, Erik K. St. Louis, Stuart J. McCarter, Silvia Rios Romenets, Amélie Pelletier, Mahmoud Cherif, et al., "Risk Factor Profile in Parkinson's Disease Subtype with REM Sleep Behavior Disorder," *Journal of Parkinson's Disease* 6, no. 1 (2016): 231–37, doi: 10.3233/JPD-150725.

38. Horne, "Human REM Sleep."

39. J. Christopher Ehlen, Allison J. Brager, Julie Baggs, Lennisha Pinckney, Chloe L. Gray, Jason P. Debruyne, Karyn A. Esser, Joseph S. Takahashi, and Ketema N. Paul, "Bmalı Function in Skeletal Muscle Regulates Sleep," *eLIFE* (2017), doi: 10.7554/ elife.26557.

40. Ibid.

41. Muhammed Ali, *The Greatest: My Own Story* (New York: Random House, 1975).

42. David Remnick, "The Outsized Life of Muhammad Ali," *New Yorker* (January 2017).

43. Kareem Abdul-Jabbar, "Kareem Abdul-Jabbar: By the Book," *New York Times*, June 4, 2017.

44. Jonathan Eig, *Ali: A Life* (New York: Houghton Mifflin, 2017).

45. Jesse Mez, Daniel H. Daneshvar, Bobak Abdolmohammadi, Alicia S. Chua, Michael L. Alosco, Patrick T. Kiernan, Laney Evers, et al., "Duration of American Football Play and Chronic Traumatic Encephalopathy," *Annals of Neurology* 87, no. 1 (2017): 116–31, doi: 10.1002/ana.25611.

46. Michael L Alosco, Yorghos Tripodis, Johnny Jarnagin, Christine M. Baugh, Brett Martin, Christine E. Chaisson, Nate Estochen, et al., "Repetitive Head Impact Exposure and Later-Life Plasma Total Tau in Former National Football League Players," *Alzheimer's and Dementia* 7 (2016): 33–40, doi: 10.1016/j.dadm.2016.11.003.

47. Xue Zhang, Fei Gao, Dongdong Wang, Chao Li., Yi Fu, Wei He, and Jianmin Zhang, "Tau Pathology in Parkinson's Disease," *Frontiers in Neurology* 9 (2018): 809, doi: 10.3389/ fneur.2018.00809.

48. Scott Gottlieb, "Head Injury Doubles Risk of Alzheimer's Disease," *British Medical Journal* 321 (2000): 1100, PMC1173459.

49. David A. Bennett, Julia A. Schneider, Zoe Arvanitakis, and Robert S. Wilson, "Overview and Findings from the Religious Orders Study," *Current Alzheimer Research* 9 (2012): 628–45, doi: 10.2174/156720512801322573.

50. Ibid.; Samuel M. Goldman, Freya Kamel, G. Webster Ross, Sarah A. Jewell, Grace S. Bhudhikanok, David Umbach, Connie Marras, et al., "Head Injury, Alpha-Synuclein Repı and Parkinson's Disease," *Annals of Neurology* 71, no. 1 (2012): 40–48, doi: 10.1002/ana.22499.

51. Slavash Jafari, Mahyar Etminan, Farhad Aminzadeh, and Ali Samil, "Head Injury and Risk of Parkinson's Disease: A Systematic Review and Meta-Analysis," *Movement Disorder* 28 (2013): 1222–29, doi: 10.1002/mds.25458.

52. Caroline M. Tanner, Freya Kamel, G. Webster Ross, Jane A. Hoppin, Samuel M. Goldman, Monica Korell, Connie Maras, et al., "Rotenone, Paraquat and Parkinson's Disease," *Environmental Health Perspectives* 119, no. 6 (2011): 866–72, doi: 10.1289/ehp.1002839.

53. J. William Langston, "The MPTP Story," *Journal of Parkinson's Disease* 7 (suppl 1): S11–19, doi: 10.3233/JPD-179006.

54. Arvid Carlsson and Maria L. Carlsson, "A Dopaminergic Deficit Hypothesis of Schizophrenia: The Path to Discovery," *Dialogues in Clinical Neuroscience* 8, no. 1 (2006): 137–42, doi: 10.31887/DCNS.2006.8.1/acarlsson.

55. Ibid.

56. Robert L. Nussbaum, "The Identification of Alpha-Synuclein as the First Parkinson Disease Gene," *Journal of Parkinson Disease* 7, suppl 1 (2017): S43–49, doi: 10.3233/JPD-179003.

57. Thomas G. Beach, Charles H. Adler, Lucia I. Sue, Linda Vedders, Lihfen Lue, Charles L. White III, Haru Akiyama, et al., "Multi-organ Distribution of Phosphorylated Alpha-Synuclein Histopathology in Subjects with Lewy Body Disorders," *Acta Neuropathology* 119, no. 6 (2010): 689–702, doi: 10.1007/s00401-010-0664-3.

58. Christine A. Klein and Ana Westenberger, "Genetics of Parkinson's Disease," *Cold Spring Harbor Perspectives on Medicine* 2, no. 1 (2012): a008888, doi: 10.1101/cshperspect.a008888.

59. Lori K. Smith, Nafisa M. Jadevji, Keri L. Colwell, S. Katrina Perehudoff, and Gerlinde A. Metz, "Stress Accelerates Neural Degeneration and Exaggerates Motor Symptoms in a Rat Model of Parkinson's Disease," *European Journal of Neuroscience* 27, no. 8 (2008): 2133–46, doi: 10.1111/j.1460-9568.2008.06177.x.

60. Valentina Oppo, Marta Melis, Melania Melis, Iole Tomassini Barbarossa, and Giovanni Cossu, "'Smelling and Tasting' Parkinson's Disease: Using Senses to Improve the Knowledge of Disease," *Frontiers in Aging Neuroscience* 12 (2020): 43, doi: 10.3389/fnagi.2020.00043.

5. More Than a Glue

Epigraph: Nicola Allen and Ben Barres, "Glia—More Than Just Brain Glue," *Nature* 457 (2009): 677, doi: 10.1038/457675a.

1. Ferris Jabr, "Know Your Neurons: What Is the Ratio of Glia to Neurons in the Brain?" *Brainwaves* (blog), *Scientific American*, June 13, 2012, https://blogs.scientificamerican.com/brainwaves/.

2. Kanav Bhatheja and Jeffrey Field, "Schwann Cells: Origins and Role in Axonal Maintenance and Regeneration," *International Journal of Biochemistry & Cell Biology* 38, no. 12 (2006): 1995–99, doi: 10.1016/j.biocel.2006.05.007.

3. Lorena Arancibia-Cárcamo, Marc C. Ford, Lee Cossell, Kinji Ishida, Koujiro Tohyama, and David Attwell, "Node of Ranvier Length as a Potential Regulator of Myelinated Axon Conduction Speed," *eLife* 6 (2017): e23329, doi: 10.7554/eLife.23329.

4. M. W. Hess, E. Kirschning, K. Pfaller, P. L. Debbage, H. Hohenberg, and G. Klima, "5000-Year-Old Myelin: Uniquely Intact in Molecular Configuration and Fine Structure," *Current Biology* 8, no. 15 (1998): R512–13, doi: 10.1016/s0960-9822(07)00334-x.

5. Albert R. Zink and Frank Maixner, "The Current Situation of the Tyrolean Iceman," *Gerontology* 65, no. 6 (2019): 699–706, doi: 10.1159/000501878.

6. Nicolas Snaidero and Mikael Simons, "Myelination at a Glance," *Journal of Cell Science* 127 (2014): 2999–3004, doi: 10.1242/jcs.151043.

7. Klaus Armin Nave and Hauke B. Werner, "Myelination of the Nervous System: Mechanisms and Functions," *Annual Review of Cell Developmental Biology* 30 (2014): 503–33, doi: 10.1146/annurev-cellbio-100913-013101; Thomas Philips and Jeffrey D. Rothstein, "Oligodendroglia: Metabolic Supporters of Neurons," *Journal of Clinical Investigation* 127, no. 9 (2017): 3271–80, doi: 10.1172/JCI90610.

8. J. L. Salzer and B. Zalc, "Myelination," *Current Biology* 26, no. 20 (2016): R937–80, doi: 10.1016/j.cub.2016.07.074.

9. Virginia García-Martín, Pablo García-López, and Miguel Freire, "Cajal's Contributions to Glia Research," *Trends in Neuroscience* 30, no. 9 (2007): 479–87, doi: 10.1016/j.tins.2007.06.008.

10. Fernando Peréz-Cerdá, María Victoria Sánchez-Gómez, and Carlos Matute, "Pío del Río Hortega and the Discovery of the Oligodendrocytes," *Frontiers in Neuroanatomy* 9 (2015): 92, doi: 10.3389/fnana.2015.00092.

11. Ibid.

12. Salzer and Zalc, "Myelination."

13. Chi-Hua Yu, Zhao Qin, Francisco J. Martin-Martinez, and Markus J. Buehler, "A Self-Consistent Sonification Method to Translate Amino Acid Sequences into Musical Compositions and Application in Protein Design Using Artificial Intelligence," *American Chemical Society* 13, no. 7 (2019), doi: 10.1021/acsnano.9b02180.

14. Anna Isabelle Boullerne, "The History of Myelin," *Experimental Neurology* 283 (2016): 431–45, doi: 10.1016/j.expneurol.2016.06.005.

15. Michael W. Salter and Beth Stevens, "Microglia Emerge as Central Players in Brain Disease," *Nature Medicine* 23, no. 9 (2017): 1018–27, doi: 10.1038/nm.4397.

16. Amanda Brosius Lutz and Ben A. Barres, "Contrasting the Glial Response to Axon Injury in the Central and Peripheral Nervous System," *Developmental Cell* 28, no. 1 (2014): 7–17, doi: 10.1016/j.devcel.2013.12.002; Susanne A. Wolf, H. W. G. M. Boddeke, and Helmut Kettenmann, "Microglia in Physiology and Disease," *Annual Review of Physiology* 79 (2017): 619–43, doi: 10.1146/annurev-physiol-022516-034406.

17. Florent Ginhoux, Melanie Greter, Marlene Leboeuf, Sayan Nandi, Peter Seem, Solen Gokhan, Mark F. Mehler, et al., "Fate Mapping Analysis Reveals that Adult Microglia Derive from Primitive Macrophages," *Science* 330, no. 6005 (2010): 841–45, doi: 10.1126/science.1194637.

18. Gustavo M. de Almeida, Francisco M. B. Germiniani, and Hélio A. G. Teive, "The Seminal Role Played by Pierre Marie in Neurology and Internal Medicine," *Arquivos de Neuro-psiquiatria* 73, no. 10 (2015): 887–89, doi: 10.1590/0004-282X20150126.

19. Jonathan Morena, Anirudh Gupta, and J. Chad Hoyle, "Charcot-Marie-Tooth: From Molecules to Therapy," *International Journal of Molecular Science* 20, no. 14 (2019): 3419. doi: 10.3390/ijms20143419.

20. Ibid.

21. Umesh Kalane, Chaitanya Datar, and Anita Mahadevan, "First Reported Case of Charcot Marie Tooth Disease Type 4C in a Child from India with SH3TC2 Mutation but Absent Spinal Deformities," *Neurology India* 63, no. 1 (2015): 395–98, https://go.gale.com/ps/i.do?p=AONE&u=googlescholar&id=GALE|A417460153&v=2.1&it=r&sid=AONE&asid=f0ab86c7n.

22. Samuel David and Albert J. Aguayo, "Axonal Elongation into Peripheral Nervous System 'Bridges' after Central Nervous System Injury in Adult Rats," *Science* 214, no. 4523 (1981): 931–33, doi: 10.1126/science.6171034.

23. Paul H. Patterson, "On the Importance of Being Inhibited, or Saying No to Growth Cones," *Neuron* 1 (1988): 261–67, doi: 10.1016/0896-6273(88)90074-8.

24. Antonio Schmandke, Andre Schmandke, and Martin E. Schwab, "Nogo-A: Multiple Roles in CNS Development, Maintenance, and Disease," *Neuroscientist* 20, no. 4 (2014): 372–86, doi: 10.1177/1073858413516800.

25. Ibid.

26. Omar de Faria, Ewa Anastazia, Claudai Pama, Kimberley Evans, Aryna Luzhnskkaya, and Ragnhildur Thóra Káradóttir, "Neuroglial Interactions Underpinning Myelin Plasticity," *Developmental Neurobiology* 78, no. 2 (2018): 93–107, doi: 10.1002/dneu.225339.

27. Erin M. Gibson, Anna C. Geraghty, and Michelle Monje, "Bad Wrap: Myelin and Myelin Plasticity in Health and Disease," *Developmental Neurobiology* 78, no. 2 (2018): 123–35, doi: 10.1002/dneu.2254.

28. Martin Kaag Rasmussen, Humberto Mestre, and Maiken Nedergaard, "The Glymphatic Pathway in Neurological Disorders," *Lancet Neurology* 17, no. 11 (2018): 1016–24, doi: 10.1016/S1474-4422(18)30318-1.

29. Lulu Xie, Hongyi Kang, Qiwu Xu, Michael J. Chen, Yonghong Liao, Meenakshisundaram Thiyagarajan, John O'Donnell, et al., "Sleep Drives Metabolite Clearance from the Adult Brain," *Science* 342, no. 6156 (2013), doi: 10.1126/science.

30. Rasmussen, Mestre, and Nedergaard, "The Glymphatic Pathway."

31. Cynthia C. Souza, Camila Hirotsu, Eduardo L. A. Neves, Lidiane C. L. Santos, Iandra M. P. F. Costa, Catarina A. Garcez, Paula S. Nunes, and Adriano Antunes, "Sleep Pattern in Charcot-Marie-Tooth Disease Type 2: Report of Family Case Series," *Journal of Clinical Sleep Medicine* 11, no. 3 (2015): 205–11, doi: 10.5664/jcsm.4526.

32. Ibid.

33. Schmandke, Schmandke, and Schwab, "Nogo-A"; de Faria et al., "Neuroglial Interactions Underpinning Myelin Plasticity."

6. A Lack of Tears

Epigraph: Conrad M. Riley, Richard L. Day, David MCL. Greeley, and William S. Langford, "Central Autonomic Dysfunction with Defective Lacrimation: Report of Five Cases," *Pediatrics* 3, no. 4 (1949): 468.

1. Riley et al., "Central Autonomic Dysfunction with Defective Lacrimation."

2. Kelly Freeman, David S. Goldstein, and Charles R. Thompson, *The Dysautonomia Project. Understanding Autonomic Nervous System Disorders for Physicians and Patients* (Sarasota, FL: Bardolf, 2015).

3. Max J. Hilz, Felicia B. Axelrod, Andreas Bickel, Brigitte Stemper, Miroslaw Brys, Gwen Wendelschafer-Crabb, and William R. Kennedy, "Assessing Function and Pathology in Familial Dysautonomia: Assessment of Temperature, Perception, Sweating and Cutaneous Innervation," *Brain* 127 (2004): 2090–98, doi: 10.1093/brain/awh235.

4. Lucy Norcliffe-Kaufmann, Susan A. Slaugenhaupt, and Horacio Kaufmann, "Familial Dysautonomia: History, Genotype, Phenotype and Translational Research," *Progress in Neurobiology* 152 (2017): 131–48, doi: 10.1016/j.pneurobio.2016.06.003.

5. Sigal Portnoy, Channa Maayan, Jeanna Tsenter, Yonah Ofran, Vladimir Goldman, Nurit Hiller, Naama Karniel, and Isabella Schwartz, "Characteristics of Ataxic Gait in Familial Dysautonomia," *PLoS One* 13, no. 4 (2018): e0196599, doi: 10.1371/journal.pone.0196599.

6. Vaughan G. Macefield, Lucy Norcliffe-Kaufmann, Joel Gutiérrez, Felicia B. Axelrod, and Horacio Kaufmann, "Can Loss of Muscle Spindle Afferents Explain the Ataxic Gait in Riley-Day Syndrome," *Brain* 134, no. 11 (2011): 3198–3208, doi: 10.1093/brain/awr168.

7. Freeman, Goldstein, and Thompson, *The Dysautonomia Project.*

8. Jose-Alberto Palma, Alex Gileles-Hillel, Lucy Norcliffe-Kaufmann, and Horacio Kaufmann, "Chemoreflex Failure and Sleep-Disordered Breathing in Familial Dysautonomia: Implications for Sudden Death during Sleep," *Autonomic Neuroscience* 218 (2019): 10–15, doi: 10.1016/j.autneu.2019.02.003,

9. Max J. Hilz, Sebastian Moeller, Susanne Buechner, Hanna Czarkowska, Indu Ayappa, Felicia B. Axelrod, and David M. Rapoport, "Obstructive Sleep-Disordered Breathing Is More Common than Central in Mild Familial Dysautonomia," *Journal of Clinical Sleep Medicine* 12, no. 12 (2016): 1607–14, doi: 10.5664/ jcsm.6342.

10. Kanwalijit Singh, Jose-Alberto Palma, Horacio Kaufmann, Nataliya Tkachenko, Lucy Norcliffe-Kaufmann, Christy Spalink, Mikhail Kazachkov, and Sanjeev V. Kothare, "Prevalence and Characteristics of Sleep-Disordered Breathing in Familial Dysautonomia," *Sleep Medicine* 45 (2018): 33–38, doi: 10.1016/ j.sleep. 2017.12.013.

11. Jose-Alberto Palma, Christy Spalink, Erin P. Barnes, Lucy Norcliffe-Kaufmann, and Horacio Kaufmann, "Neurogenic Dysphagia with Undigested Macaroni and Megaesophagus in Familial Dysautonomia," *Clinical Autonomic Research* 28, no. 1 (2018): 125–26, doi: 10.1007/ s10286-017-0487-6.

12. Lucy Norcliffe-Kaufmann, Susan A. Slaugenhaupt, and Horacio Kaufmann, "Familial Dysautonomia: History, Genotype, Phenotype and Translational Research," *Progress in Neurobiology* 152 (2017): 131–48, doi: 10.1016/j.pneurobio.2016.06.003.

13. J. E. Hall, "Control of Blood Pressure by the Renin-Angiotensin-Aldosterone System," *Clinical Cardiology* 14, no. 8 (1991): IV6–21, doi: 10.1002/clc.4960141802.

14. Lior Elkayam, Albert Matalon, Chi-Hong Tseng, and Felicia Axelrod, "Prevalence and Severity of Renal Disease in Familial Dysautonomia," *American Journal of Kidney Disease* 48, no. 5 (2006): 780–86, doi: 10.1053/j.ajkd.2006.07.024.

15. Horacio Kaufmann and Roger Hainsworth, "Why Do We Faint?" *Muscle & Nerve* 24, no. 8 (2001): 981–83, doi: 10.1002/mus.1102.

16. A. J. Aguayo, P. V. Cherunada, M. B. Nair, and G. M. Bray, "Peripheral Nerve Abnormalities in the Riley-Day Syndrome," *Neurology* 24 (1971): 106–16, doi: 10.1001/archneur.1971.00480320034003.

17. Ibid.

18. Jose-Alberto Palma, Lucy Norcliffe-Kaufmann, Cristina Fuente-Mora, Leila Percival, Carlos Mendoza-Santiesteban, and Horacio Kaufmann, "Current Treatments in Familial Dysautonomia," *Expert Opinion on Pharmacotherapy* 15, no. 10 (2014): 2633–71, doi: 10.1517/14656566.2014.970530.

19. David S. Feldman, David E. Ruchelsman, Daniel B. Spencer, Joseph J. Straight, Mark E. Schweitzer, and Felicia Axelrod, "Peripheral Arthropathy in Hereditary Sensory and Autonomic Neuropathy Types III and IV," *Journal of Pediatric Orthopedics* 29, no. 1 (2009): 91–97, doi: 10.1097/BPO.0b013e31818f9cc4; David S. Goldstein, "Dysautonomia in Parkinson's Disease: Neurocardiological Abnormalities," *Lancet Neurology* 2, no. 11 (2013): 669–76, doi: 10.1016/s1474-4422(03)00555-6.

20. Sophie Pezet and Stephen B. McMahon, "Neurotrophins: Mediators and Modulators of Pain," *Annual Review of Neuroscience* 29 (2006): 507–38, doi: 10.1146/annurev.neuro.29.051605.112929.

21. S. L. Anderson, R. Coli, I. W. Daly, E. A. Kichula, M. J. Rork, S. A. Volpi, J. Ekstein, and B. Y. Rubin, "Familial Dysautonomia Is Caused by Mutations of the IKAP Gene," *American Journal of Human Genetics* 68, no. 3 (2001): 753–58, doi: 10.1086/318808.

22. Ibid.

23. Anat Blumenfeld, Susan A. Slaugenhaupt, Felicia B. Axelrod, Diane E. Lucente, Channa Maayan, Christopher B. Liebert, Lauria J. Ozelius, et al., "Localization of the Gene for Familial Dysautonomia on Chromosome 9 and Definition of DNA Markers for Genetic Diagnosis," *Nature Genetics* 4 (1993): 160–64, doi: 10.1038/ng0693-160.

24. Anderson et al., "Familial Dysautonomia Is Caused by Mutations of the IKAP Gene."

25. Math P. Cuajungco, Maire Leyne, James Mull, Sandra P. Gill, Weining Lu, David Zagzag, Felicia B. Axelrod, Channa Maayan, et al., "Tissue-Specific Reduction in Splicing Efficiency of IKBKAP Due to the Major Mutation Associated with Familial Dysautonomia," *American Journal of Human Genetics* 72, no. 3 (2009): 749–58, doi: 10.1086/368263.

26. Anderson et al., "Familial Dysautonomia"; Horacio Kaufman, Kristy Nahm, Dushyant Purohit, and David Wolfe, "Autonomic Failure as the Initial Presentation of Parkinson Disease and Dementia with Lewy Bodies," *Neurology* 63, no. 6 (2004): 1093–95, doi: 10.1212/01.wnl.0000138500.73671.dc.

27. Felicia B. Axelrod and Horacio Kaufmann, "Hereditary Sensory and Autonomic Neuropathies," in *Neuromuscular Disorders of Infancy, Childhood and Adolescence: A Clinician's Approach*, 2nd ed., ed. Basil T. Darras, H. Roydan Jones Jr., Monique M. Ryan, and Darryl C. DeVivo (New York: Elsevier), 340–52; Coreen Schwartzlow and Mohamed Kazamel, "Hereditary Sensory and Autonomic Neuropathies: Adding More to the Classification," *Current Neurology and Neuroscience Reports* 19, no. 8 (2019): 52, doi: 10.1007/s11910-019-0974-3.

28. Yimin Hua, Kentaro Sahashi, Frank Rigo, Gene Hung, Guy Horev, C. Frank Bennett, and Adrian R. Krainer, "Peripheral SMN Restoration Is Essential for Long-Term Rescue of a Severe Muscular Atrophy Mouse Model," *Nature* 478: 123–26, doi: 10.1038/nature10485.

29. Rahul Sinha, Young Jin Kim, Tomoki Nomakuchi, Kentaro Sahashi, Yimin Hua, Frank Rigo, C. Frank Bennett, and Adrian R. Krainer, "Antisense Oligonucleotides Correct the Familial Dysautonomia Splicing Defect in IKBKAP Transgenic Mice," *Nucleic Acids Research* 46, no. 10 (2018): 4833–44, doi: 10.1093/nar/gky249.

30. Michaela Auer-Grumbach, "Hereditary Sensory Neuropathy Type I," *Orphanet Journal of Rare Diseases* 3, doi: 10.1186/1750-1172-3-7.

31. Felicia B. Axelrod and Gabrielle Gold-von Simson, "Hereditary Sensory and Autonomic Neuropathies: Types II, III and IV," *Orphanet Journal of Rare Diseases* (2007): 39, doi: 10.1186/1750-1172-2-39.

32. Yashido Indo, "NGF-dependent Neurons and Neurobiology. Of Emotions and Feelings: Lessons from Congenital Insensitivity to Pain with Anhidrosis," *Neuroscience Biobehavioral Reviews* 87 (2018): 1–16, doi: 10.1016/j.neubiorev.2018.01.013.

33. Ibid.

34. Giovanna Testa, Marco Mainardi, Chiara Morelli, Francesco Olimpico, Laura Pancrazi, Carla Petrella, Cinzia Severini, et al., "The NGF R100W Mutation Specifically Impairs Nociception without Affecting Cognitive Performance in a Mouse Model of Hereditary Sensory and Autonomic Neuropathy Type V," *Journal of Neuroscience* 39, no. 49 (2019): 9702–971, doi: 10.1523/JNEUROSCI.0688-19.2019.

7. The Power of Touch

Epigraph: Temple Grandin and Richard Panek, *The Autistic Brain: Helping Different Kinds of Minds Succeed* (New York: Houghton Mifflin Harcourt, 2013), 70.

1. Jin Baio, Lisa Wiggins, Deborah L. Christensen, Matthew J. Maenner, Julie Daniels, Zachary Warren, Margaret Kurzius-Spencer, et al., "Prevalence of Autism Spectrum Disorder among Children Aged 8 Years," Centers for Disease Control and Prevention *MMWR Surveillance Summary 2018* 67, no. 6 (2019): 1–23, doi: 10.15585/mmwr.ss6706a1.

2. Abbe V. Kirby, Brian A. Boyd, Kathryn L. Williams, Richard A. Faldowski, and Grace T. Baranek, "Sensory and Repetitive Behaviors among Children with Autism," *Autism* 21, no. 2 (2017): 142–54, doi: 10.1177/1362361316632710.

3. Sungji Ha, In-Jung Sohn, Namwook Kim, Hyeon Jeong Sim, and Keun-Ah Cheon, "Characteristic of Brains in Autism Spectrum Disorder: Structure, Function and Connectivity across the Lifespan," *Experimental Neurobiology* 24, no. 4 (2015): 273–84, doi: 10.5607/en.2015.24.4.273.

4. Rachel Loomes, Laura Hull, and William Polmear Locke Mandy, "What Is the Male-to-Female Ratio in Autism Spectrum Disorder? A Systematic Review and Meta-Analysis," *Journal of the American Academy of Child and Adolescent Psychiatry* 56, no. 6 (2017): 466–74, doi: 10.1016/j.jaac.2017.03.013.

5. Andrew Scull, *Hysteria: The Disturbing History* (New York: Oxford University Press, 2009).

6. Bonnie Evans, "How Autism Became Autism: The Radical Transformation of a Central Concept of Child Development in Britain," *History of Human Sciences* 26, no. 3 (2013): 3–31, doi: 10.1177/0952695113484320.

7. Leo Kanner, "Autistic Disturbances of Affective Contact," *Nervous Child* 2 (1943): 217–50.

8. Ibid.

9. Ibid.

10. J. B. Barahona-Corrêa and Carlos N. Filipe, "A Concise History of Asperger Syndrome: The Short Reign of a Troublesome Diagnosis," *Frontiers in Psychology* 6 (2016): 2024, doi 10.3389/fpsyg.2015.02024.

11. Edith Sheffer, *Asperger's Children: The Origins of Autism in Nazi Vienna* (New York: W. W. Norton, 2018).

12. Herwig Czech, "Hans Asperger, National Socialism, and 'Race Hygiene' in Nazi-Era Vienna," *Molecular Autism* 9 (2018): 29, doi: 10.1186/s13229-018-0208-6.

13. Fred R. Volkmar and James C. McPartland, "From Kanner to DSM-5: Autism as an Evolving Diagnostic Concept," *Annual Review of Clinical Psychology* 10 (2014): 193–212, doi: 10.1146/annurev-clinpsy-032813-153710.

14. Zhen Zheng, Peng Zheng, and Xiaobing Zou, "Association between Schizophrenia and Autism Spectrum Disorder: A Systematic Review and Meta-Analysis," *Autism Research* 11, no. 8 (2018): 1110–19, doi: 10.1002/aur.1977.

15. Melissa D. Thye, Haley M. Bednarz, Abbey J. Herringshaw, and Emma B. Sartin, "The Impact of Atypical Sensory Processing on Social Impairments in Autism Spectrum Disorder," *Developmental Cognitive Neuroscience* 29 (2018): 151–67, doi: 10.1016/j.dcn.2017.04.010.

16. B. Birmaher, D. A. Brent, L. Chiappetta, J. Bridge, S. Monga, and M. Baugher, "Psychometric Properties of the Screen for Child Anxiety Related Emotional Disorders (SCARED): A Replication Study," *Journal of the American Academy of Child and Adolescent Psychiatry* 38 (1999): 1230–36, doi: 10.1097/00004583-199910000-00011.

17. Ibid.

18. Mark Haddon, *The Curious Incident of the Dog in the Night-Time* (New York: Vintage, 2003).

19. F. Kyle Satterstrom, Jack A. Kosmicki, Jiebiao Wang, Michael S. Breen, Silvia De Rubeis, Joon-Yong An, Minshi Peng, et al., "Large-Scale Exome Sequencing Study Implicates Both Developmental and Functional Changes in the Neurobiology of Autism," *Cell* 180, no. 3 (2020): 568–84, 10.1016/j.cell.2019.12.036.

20. Laura J. Wright, "Analysis of Sequences Pegs 102 Top Autism Genes," *Spectrum News*, November 2, 2020, www.spectrumnews.org/news/analysis.

21. Paul Nurse, "Biology Must Generate Ideas as well as Data," *Nature* 597, no. 7876 (2021): 305, doi: 10.1038/d41586-021-02480-z.

22. Charles A. Nelson, "Romanian Orphans Reveal Clues to Origins of Autism," *Washington Post*, May 14, 2017.

23. April R. Levin, Nathan A. Fox, Charles H. Zeanah, and Charles A. Nelson, "Social Communication Difficulties and Autism in Previously Institutionalized Children," *Journal*

of the American Academy of Child and Adolescent Psychiatry 54, no. 2 (2015): 108–15, doi: 10.1016/j.jaac.2014.11.011.

24. Edmund J. S. Sonuga-Barke, Mark Kennedy, Ronert Kumsta, Nicky Knights, Dennis Golm, Michael Rutter, Barbara Maughan, Wolff Schlotz, and Jana Kreppner, "Child to Adult Neurodevelopmental and Mental Health Trajectories after Early Life Deprivation: The Young Adult Follow-Up of the Longitudinal English and Romanian Adoptees Study," *Lancet* 369, no. 10078 (2017): 1539–48, doi: 10.1016/S0140-6736(17)30045-4.

25. Clare S. Allely, "Pain Sensitivity and Observer Perception of Pain in Individuals with Autistic Spectrum Disorder," *Scientific World Journal* 13 (2013): 91616. doi: 1155/32013/916178.

26. Temple T. Grandin, "Visual Thinking, Sensory Problems and Communication Difficulties," Synapse.org, www.autism-help.org/story-sensory-communication.htm.

27. Scott D. Tomchek and Winnie Dunn, "Sensory Processing in Children with and without Autism: A Comparative Study Using the Short Sensory Profile," *American Journal of Occupational Therapy* 61, no. 2 (2007): 190–200, doi: 10.5014/ajot.61.2.190; Teresa Tavassoli, Lucy J. Miller, Sarah A. Schoen, Darci M. Nielsen, and Simon Baron-Cohen, "Sensory Over-Responsivity in Adults with Autism Spectrum Conditions," *Autism* 18, no. 4 (2014): 428–32, doi: 10.1177/1362361313477246.

28. Caroline E. Robertson and Simon Baron-Cohen, "Sensory Perception in Autism," *Nature Reviews of Neuroscience* 18, no. 11 (2017): 84, doi: 10.1038/nrn.2017.112.

29. Brian A. Boyd, Grace T. Baranek, John Sideris, Michele D. Poe, Linda R. Watson, Elena Patten, and Heather Miller, "Sensory Features and Repetitive Behaviors in Children with Autism and Developmental Delays," *Autism Research* 3, no. 2 (2010): 78–87, doi: 10.1002/aur.124.

30. Carissa Cascio, Francis McGlone, Stephen Folger, Vinay Tannan, Grace Baranek, Kevin A. Pelphrey, and Gregory Essick, "Tactile Perception in Adults with Autism: A Multidimensional Psychophysical Study," *Journal of Autism and Developmental Disorders* 38, no. 1 (2008): 127–37, doi: 10.1007/s10803-007-0370-8; Carissa Cascio, David Moore, and Francis McGlone, "Social Touch and Human Development," *Developmental Cognitive Neuroscience* 35 (2019): 5–11, doi: 10.1016/j.dcn.2018.04.009.

31. Cascio, Moore, and McGlone, "Social Touch and Human Development."

32. Nicci Grace, Beth P. Johnson, Nicole J. Rinehart, and Peter G. Enticott, "Are Motor Control and Regulation Problems Part of the ASD Motor Profile? A Handwriting Study," *Developmental Neuropsychology* 43, no. 7 (2018): 581–94, doi: 10.1080/87565641.2018.1504948.

33. Christina T. Fuentes, Stewart H. Mostofsky, and Amy J. Bastian, "Children with Autism Show Specific Handwriting Impairments," *Neurology* 73, no. 19 (2009): 1532–37, doi: 10.1212/WNL.0b013e3181c0d48c; Grace et al., "Are Motor Control and Regulation Problems Part of the ASD Motor Profile?"

34. India Morrison, Line S. Loken, and Håkan Olausson, "The Skin as a Social Organ," *Experimental Brain Research* 204, no. 3 (2010): 305–14, doi: 10.1007/s00221-009-2007-y.

35. Grace T. Baranek, Linda R. Watson, Brian A. Boyd, Michele D. Poe, Fabian J. David, and Lorin McGuire, "Hyporesponsiveness to Social and Nonsocial Sensory Stimuli in Children with Autism, Children with Developmental Delays and Typically Developing

Children," *Developmental Psychopathology* 25, no. 2 (2013): 307–20, doi: 10.1017/S0954579412 001071.

36. Stephen M. Maricich, Scott A. Wellnitz, Aislyn M. Nelson, Daine R. Lesniak, Gregory J. Gerling, Ellen A. Lumpkin, and Huda Y. Zoghbi, "Merkel Cells Are Essential for Light Touch Responses," *Science* 324, no. 5934 (2009): 1580–82, doi: 10.1126/science.1172890.

37. Sophie Pezet and Stephen B. McMahon, "Neurotrophins: Mediators and Modulators of Pain," *Annual Review of Neuroscience* 29 (2006): 507–38, doi: 10.1146/annurev.neuro.29.051605.112929.

38. Oliver Sacks, *The Man Who Mistook His Wife for a Hat and Other Clinical Tales* (New York: Summit Books, 2006).

39. Laurel L. Orefice, Amanda L. Zimmerman, Anda M. Chirila, Steven J. Sleboda, Joshua P. Head, and David D. Ginty, "Peripheral Mechanosensory Neuron Dysfunction Underlies Tactile and Behavioral Deficits in Mouse Models of ASDs," *Cell* 166, no. 2 (2016): 299–313, doi: 10.1016/j.cell.2016.05.033.

40. Alexander H. Tuttle, Victoria B. Bartsch, and Mark J. Zylka, "The Troubled Touch of Autism," *Cell* 166, no. 2 (2016): 273–74, doi: 10.1016/j.cell.2016.06.054.

41. Matthew J. Hertenstein, Julia M. Verkamp, Alyssa M. Kerestes, and Rachael M. Holmes, "The Communicative Functions of Touch in Humans, Nonhuman Primates, and Rats: A Review and Synthesis of the Empirical Research," *Genetic, Social, and General Psychology Monographs* 132, no. 1 (2006): 5–94, doi: 10.3200/mono.132.1.5-94.

42. David R. Simmons, Ashley E. Robertson, Lawrie S. McKay, Erin Toal, Phil McAleer, and Frank E. Pollick, "Vision in Autism Spectrum Disorders," *Vision Research* 49, no. 22 (2009): 2705–39, doi: 10.1016/j.visres.2009.08.005; Liron Rozenkrantz, Ditza Zachor, Iris Heller, Anton Plotkin, Aharon Weissbrod, Kobi Snitz, Lavi Secundo, and Noam Sobel, "A Mechanistic Link between Olfaction and Autism Spectrum Disorder," *Current Biology* 25, no. 14 (2015): 1904–10, doi: 10.1016/j.cub.2015.05.048; T. Tavassoli and S. Baron-Cohen, "Taste Identification in Adults with Autism Spectrum Conditions," *Journal of Autism and Developmental Disorders* 42, no. 7 (2012): 1419–24, doi: 10.1007/s10803-011-1377-8.

43. Rozenkrantz et al., "A Mechanistic Link between Olfaction and Autism Spectrum Disorder."

44. Nicolaas A. J. Puts, Ericka L. Wodka, Mark Tommerdahl, Stewart H. Mostofsky, and Richard A. E. Edden, "Impaired Tactile Processing in Children with Autism Spectrum Disorder," *Journal of Neurophysiology* 111, no. 9 (2014): 1803–11, doi: 10.1152/jn.00890.2013.

45. Jonathan Kipnis, "Immune System: The 'Seventh Sense,'" *Journal of Experimental Medicine* 215, no. 2 (2018): 397–98, doi.org/10.1084/jem.20172295.

46. Hjördís Osk Atladóttir, Poul Thorsen, Diana E. Schendel, Lars Østergaard, Saane Lemcke, and Erik T. Parner, "Association of Hospitalization for Infection in Childhood with Diagnosis of Autism Spectrum Disorders," *Archives of Pediatric Adolescent Medicine* 164, no. 5 (2010): 470–77, doi: 10.1001/archpediatrics.2010.9.

47. Amani Alharthi, Safiah Alhazmi, Najla Alburae, and Ahmed Bahieldin, "The Human Gut Microbiome as a Potential Factor in Autism Spectrum Disorder," *International Journal of Molecular Science* 23, no. 3 (2022): 1363, doi: 10.3390/ijms23031363.

48. Gloria B. Choi, Yeong S. Yim, Helen Wong, Sangdoo Kim, Hyunju Kim, Sandwon V. Kim, Charles A. Hoeffer, Dan R. Littman, and Jun R. Huh, "The Maternal Interleukin-17a Pathway in Mice Promotes Autism-Like Phenotypes in Offspring," *Science* 351, no. 6276 (2016): 933–39, doi: 10.1126/science.aad0314.

49. Natalia V. Malkova, Collin Z. Yu, Elaine Y. Hsiao, Marilyn J. Moore, and Paul H. Patterson, "Maternal Immune Activation Yields Offspring Displaying Mouse Versions of the Three Core Symptoms of Autism," *Brain, Behavior, and Immunity* 26, no. 4 (2012): 607–16, doi: 10.1016/j.bbi.2012.01.011.

50. Malkova et al., "Maternal Immune Activation."

51. Atladóttir et al., "Association of Hospitalization."

52. Vincent Du Vigneaud, "Trial of Sulfur Research: From Insulin to Oxytocin," *Science* 123, no. 3205 (1956): 967–74, doi: 10.1126/science.123.3205.967.

53. Thomas R. Insel and Terrence J. Hulihan, "A Gender-Specific Mechanism for Pair Bonding: Oxytocin and Partner Preference Formation for Monogamous Voles," *Behavioral Neuroscience* 109, no. 4 (1995): 782–89, doi: 10.1037//0735-7044.109.4.782.

54. Larry J. Young, Miranda M. Lim, Brenden Gingrich, and Thomas R. Insel, "Cellular Mechanisms of Social Attachment," *Hormones and Behavior* 40 (2001): 133–38, doi: 10.1006/hbeh.2001.1691.

55. Olga L. Lopatina, Yulia K. Komleva, Yana V Gorina, Haruhiro Higashida, and Alla B. Salmina, "Neurobiological Aspects of Face Recognition: The Role of Oxytocin," *Frontiers in Behavioral Neuroscience* 12 (2018), doi: 10.3389/fnbeh.2018.00195.

56. Bianca J. Marlin, Mariela Mitre, James A. D'Amour, Moses V. Chao, and Robert C. Froemke, "Oxytocin Enables Maternal Behavior by Balancing Cortical Inhibition," *Nature* 520 (2015): 499–504, doi: 10.1038/nature14402.

57. S. Boll, A. C. de Minas, A. Raftogianni, S. C. Herpetz, and V. Grinevich, "Oxytocin and Pain Perception: From Animal Models to Human Research," *Neuroscience* 387 (2018): 149–61, doi: 10.1016/j.neuroscience.2017.09.041.

58. Yan Tang, Diego Benusiglio, Arthur Lefevre, Louis Hilfiger, Ferdinand Althammer, Ann Blugau, Daisuke Hagiwara, et al., "Social Touch Promotes Interfemale Communication via Activation of Parvocellular Oxytocin Neurons," *Nature Neuroscience* 23, no. 9 (2020): 1125–37, doi: 10.1038/s41593-020-0674-y.

59. Marilena M. DeMayo, Yun Ju C. Song, Ian B. Hinkle, and Adam J. Guastella, "A Review of the Safety, Efficacy and Mechanisms of Delivery of Nasal Oxytocin in Children: Therapeutic Potential for Autism and Prader-Willi Syndrome, and Recommendations for Future Research," *Pediatric Drugs* 19, no. 5 (2017): 391–410, doi: 10.1007/s40272-017-0248-y; Patricia S. Churchland and Piotr Winkielman, "Modulating Social Behavior with Oxytocin: How Does It Work? What Does It Mean?" *Hormones Behavior* 61, no. 3 (2012): 392–99, doi: 10.1016/j.yhbeh.2011.12.003.

60. Tavassoli et al., "Sensory Over-Responsivity"; Robertson and Baron-Cohen, "Sensory Perception in Autism"; Boyd, Baranek, and Sideris, "Sensory Features and Repetitive Behaviors."

61. Tang et al., "Social Touch Promotes Interfemale Communication."

62. Lauren L. Orefice, Jacqueline R. Mosko, Danielle T. Morency, Michael F. Wells, Aniqa Tasnim, Shawn M. Mozeika, Mengchen Ye, et al., "Targeting Peripheral Somatosensory Neurons to Improve Tactile-Related Phenotypes in ASD Models," *Cell* 178, no. 4 (2019): 867–86, doi: 10.1016/j.cell.2019.07.024; Tuttle, Bartsch, and Zylka, "The Troubled Touch of Autism"; Boyd, Baranek, and Sideris, "Sensory Features and Repetitive Behaviors"; Cascio et al., "Tactile Perception in Adults with Autism"; Cascio, Moore, and McGlone, "Social Touch and Human Development."

8. Plasticity in the Periphery

Epigraph: William James, "The Laws of Habit," *Popular Science Monthly* (February 1887), 434.

1. Norman Doidge, *The Brain That Changes Itself: Stories of Personal Triumph from the Frontiers of Brain Science* (New York: Penguin, 2007); Ryuta Kawashima, *Train Your Brain: 60 Days to a Better Brain* (Teaneck, NJ: Kumon, 2005).

2. William James, *The Principles of Psychology*, 2 vols. (New York: Henry Holt, 1890).

3. Moheb Costandi, *Neuroplasticity* (Cambridge, MA: MIT Press, 2016).

4. Anne E. Takesian and Takeo K. Hensch, "Balancing Plasticity/Stability across Brain Development," *Progress in Brain Research* 207 (2013): 3–34, doi: 10.1016/B978-0-444-63327-9.00001-1.

5. J. Sebastian Espinosa and Michael P. Stryker, "Development and Plasticity of the Primary Visual Cortex," *Neuron* 75, no. 2 (2012): 230–49, doi: 10.1016/j.neuron.2012.06.009.

6. Melanie P. Clements, Elizabeth Byrne, Luis F. Camarillo Guerrero, Anne-Laure Cattin, Leila Zakka, Azhaar Ashraf, Jemima J. Burden, et al., "The Wound Microenvironment Reprograms Schwann Cells to Invasive Mesenchymal-Like Cells to Drive Peripheral Nerve Regeneration," *Neuron* 96, no. 1 (2017): 98–114, doi: 10.1016/j.neuron.2017.09.008.

7. Michio W. Painter, Amanda Brosius Lutz, Yung-Chih Cheng, Alban Latremoliere, Kelly Duong, Christine M. Miller, Sean Posada, et al., "Diminished Schwann Cell Repair Responses Underlie Age-Associated Impaired Axonal Regeneration," *Neuron* 83, no. 2 (2014): 331–43, doi: 10.1016/j.neuron.2014.06.016.

8. Ibid.

9. Seung-Hyun Woo, Ellen A. Lumpkin, and Ardem Patapoutian, "Merkel Cells and Neurons Keep in Touch," *Trends in Cell Biology* 25, no. 2 (2015): 74–81, doi: 10.1016/j.tcb.2014.10.003; Kara L. Marshall, Rachel C. Clary, Yoshichika Baba, Rachel L. Orlowsky, Gregory J. Gerling, and Ellen A. Lumpkin, "Touch Receptors Undergo Rapid Remodeling in Healthy Skin," *Cell Reports* 17, no. 7 (2016): 1719–27, doi: 10.1016/j.celrep.2016.10.034.

10. Woo, Lumpkin, and Patapoutian, "Merkel Cells and Neurons Keep in Touch."

11. Lawrence C. Katz, "What's Critical for the Critical Period in Visual Cortex?," *Cell* 99, no. 7 (1999): 673–76, doi: 10.1016/s0092-8674(00)81665-7.

12. Lawrence C. Katz and Manning Rubin, *Keep Your Brain Alive: 83 Neurobic Exercises to Help Prevent Memory Loss and Increase Mental Fitness* (New York: Workman, 1993).

13. A. Kimberley McAllister, Lawrence C. Katz, and Donald C. Lo, "Neurotrophins and Synaptic Plasticity," *Annual Review of Neuroscience* 22 (1999): 296–319, doi: 10.1146/annurev.neuro.22.1.295.

14. Katz and Rubin, *Keep Your Brain Alive.*

15. Ibid.

16. McAllister, Katz, and Lo, "Neurotrophins and Synaptic Plasticity."

17. Anita E. Autry and Lisa M. Monteggia, "Brain-Derived Neurotrophic Factor and Neuropsychiatric Disorders," *Pharmacological Reviews* 64, no. 2 (2012): 238–58, doi: 10.1124/pr.111.005108.

18. Carl W. Cotman, Nicole C. Berchtold, and Lori-Ann Christie, "Exercise Builds Brain Health: Key Roles of Growth Factor and Inflammation," *Trends in Neuroscience* 30, no. 9 (2007): 464–72, doi: 10.1016/j.tins.2007.06.011; Angelica Miki Stein, Thays Martins Vital Silva, Flávia Gomes de Coelho, Franciel José Arantes, José Luiz Riani Costa, Elizabeth Teodoro, and Ruth Ferreira Santos-Galduróz, "Physical Exercise, IGF-1 and Cognition: A Systematic Review of Experimental Studies in the Elderly," *Dementia & Neuropsychologia* 12, no. 2 (2018): 114–22, doi: 10.1590/1980-57642018dn12-020003.

19. Andreas Hohn, Joachim Leibrock, Karen Bailey, and Yves-Alain Barde, "Identification and Characterization of a Novel Member of the Nerve Growth Factor/Brain-Derived Neurotrophic Factor Family," *Nature* 344 (1990): 339–41, doi: 10.1038/344339a0.

20. Kristen R. Maynard, Julia L. Hill, Nicholas E. Calcaterra, Mary E. Palko, Alisha Kardian, Daniel Paredes, Mahima Sukumar, et al., "Functional Role of BDNF Production from Unique Promoters in Aggression and Serotonin Signaling," *Neuropsychopharmacology* 41, no. 8 (2016): 1943–55, doi: 10.1038/npp.2015.349; Giles S. H. Yeo, Chiao-Chien Hung, Justin Rochford, Juliette Keogh, Juliette Gray, Shoba Sivaramakrishnan, Stephen O'Rahilly, and I. Sadaf Farooqi, "A De Novo Mutation Affecting Human TrkB Associated with Severe Obesity and Developmental Delay," *Nature Neuroscience* 7, no. 11 (2004): 1187–89, doi: 10.1038/nn1336; Joan C. Han, Qing-Rong Liu, MaryPat Jones, Rebecca L. Levinn, Carolyn M. Menzie, Kyra S. Jefferson-George, Diane C. Alder-Wailes, et al., "Brain-Derived Neurotrophic Factor and Obesity in the WAGR Syndrome," *New England Journal of Medicine* 359, no. 9 (2007): 918–27, doi: 10.1056/NEJMoa0801119.

21. Baoji Xu and Xiangyang Xie, "Neurotrophic Factor Control of Satiety and Body Weight," *Nature Reviews Neuroscience* 17, no. 5 (2016), 282–92, doi: 10.1038/nrn.2016.24; Karunesh Ganguly and Mu-Ming Poo, "Activity-Dependent Neural Plasticity from Bench to Bedside," *Neuron* 80, no. 3 (2013): 729–41, doi: 10.1016/j.neuron.2013.10.028.

22. Autry and Monteggia, "Brain-Derived Neurotrophic Factor."

23. See "Orpheus Reflections: The Healing Power of Music," Orpheus Chamber Orchestra, https://orpheusnyc.org/education-community/reflections.

24. Xu and Xie, "Neurotrophic Factor Control of Satiety."

25. Michelle W. Voss, Carmen Vivar, Arthur F. Kramer, and Henriette van Praag, "Bridging Animal and Human Models of Exercise-Induced Brain Plasticity," *Trends in Cognitive Sciences* 17, no. 10 (2013): 525–44, doi: 10.1016/j.tics.2013.08.001; Helen E. Scharfman

and Rene Hen, "Neuroscience: Is More Neurogenesis Always Better?," *Science* 315, no. 5810 (2007): 336–38, doi: 10.1126/science.1138711.

26. Lauretta El Hayek, Mohamad Khalifeh, Victor Zibara, Rawad Abi Assaad, Nancy Emmanuel, Nabil Karnib, Eim El-Ghandour, et al., "Lactate Mediates the Effects of Exercise on Learning and Memory through SIRT1-Dependent Activation of Hippocampal Brain-Derives Neurotrophic Factor (BDNF)," *Journal of Neuroscience* 39, no. 13 (2019): 2369–82, doi: 10.1523/JNEUROSCI.1661-18.2019.

27. Stephanie von Holstein-Rathlou, Nicolas Caesar Petersen, and Maiken Nedergaard, "Voluntary Running Enhances Glymphatic Influx in Awake Behaving Young Mice," *Neuroscience Letters* 662 (2018): 253–58, doi: 10.1016/j.neulet.2017.10.035.

28. Olie Westheimer, Cynthia McRae, Claire Henchcliffe, Arman Fesharki, Sofya Glazman, Heather Ene, and Ivan Bodis-Wollner, "Dance for PD: A Preliminary Investigation of Effects of Motor Function and Quality of Life among Persons with Parkinson's Disease," *Journal of Neural Transmission* 122, no. 9 (2015): 1263–70, doi: 10.1007/s00702-015-1380-x.

29. Cotman, Berchtold, and Christie, "Exercise Builds Brain Health."

30. Daniel R. Cleary, Alp Ozpinar, Ahmed M. Rasian, and Andrew L. Ko, "Deep Brain Stimulation for Psychiatric Disorders: Where We Are Now," *Neurosurgical Focus* 38, no. 6 (2015): E2. doi: 10.327/2015.3FOCuS1546.

31. D. Luke Fischer, Christopher J. Kemp, Allyson Cole-Strauss, Nicole K. Polinski, Katrina L. Paumier, Jack W. Lipton, Kathy Steece-Collier, et al., "Subthalamic Nucleus Deep Brain Stimulation Employs Trkb Signaling for Neuroprotection and Functional Restoration," *Journal of Neuroscience* 37, no. 28 (2017): 6786–96, doi: 10.1523/JNEUROSCI.2060-16.2017.

32. Milan R. Dimitrijevic, Yuri Gerasimenko, and Michela M. Pinter, "Evidence for a Spinal Central Pattern Generator in Humans," *Annals of the New York Academy of Science* 860 (1998): 360–78, doi: 10.1111/j.1749-6632.1998.tb09062.x.

33. Massimo Conese, Annalucia Carbone, Elisa Beccia, and Antonella Angiolillo, "The Fountain of Youth: A Tale of Parabiosis, Stem Cells, and Rejuvenation," *Open Medicine* 12 (2017): 376–83, doi: 10.1515/med-2017-0053.

34. Joseph M. Castellano, Elizabeth D. Kirby, and Tony Wyss-Coray, "Blood-Borne Revitalization of the Aged Brain," *JAMA Neurology* 71, no. 10 (2015): 1191–94, doi: 10.1001/jamaneurol.2015.1616.

35. *JAMA*, "Paul Bert (1833–1886), Aviation Physiologist," *Journal of the American Medical Association* 211 (1979): 1849–50,

36. Conese, Beccia, and Angiolillo, "The Fountain of Youth."

37. Saul A. Villeda, Kristopher E. Plambeck, Jinte Middeldorp, Joseph M. Castellano, Kira I. Mosher, Jian Luo, Lucas K. Smith, et al., "Young Blood Reverses Age-Related Impairments in Cognitive Function and Synaptic Plasticity in Mice," *Nature Medicine* 20, no. 6 (2014): 659–63, doi: 10.1038/nm.3569; Tony Wyss-Coray, "Ageing, Neurodegeneration and Brain Rejuvenation," *Nature* 539, no. 7628 (2016): 180–86, doi: 10.1038/nature20411.

38. Yvonne Naegelin, Hayley Dingsdale, Katharina Sauberli, Sabine Schadelin, Ludwig Kappos, and Yves-Alain Barde, "Measuring and Validating the Levels of Brain-Derived Neurotrophin Factor in Human Serum," *eNeuro* 5, no. 2 (2018): doi: 10.1523/ENEURO.0419-17.2018.

39. Hans Thoenen and M. Sendtner, "Neurotrophins: From Enthusiastic Expectations through Sobering Experiences to Rational Therapeutic Approaches," *Nature Neuroscience* 5 (2002): 1046–50, doi: 10.1038/nn938.

40. John C. Newman and Eric Verdin, "Ketone Bodies as Signaling Metabolites," *Trends in Endocrine Metabolism* 25, no. 1 (2014): 42–52, doi: 10.1016/j.tem.2013.09.002; Odette Leiter, Suse Seidemann, Rupert W. Overall, Beáta Ramasz, Nicole Rund, Sonja Schallenberg, Tatyana Grinenko, et al., "Exercise-Induced Activated Platelets Increase Adult Hippocampal Precursor Proliferation and Promote Neuronal Differentiation," *Stem Cell Reports* 12, no. 4 (2019): 667–79, doi: 10.1016/j.stemcr.2019.02.009.

41. Sama F. Sleiman, Jeffrey Kessna-Henry, and R. Al-Haddad, "Exercise Promotes the Expression of BDNF through the Action of the Ketone Body β-Hydroxylbutyrate," *eLife* (2016), doi: 10.7554/eLife.15092.

42. Zarife Sahenk, Gloria Galloway, Kelly Reed Clark, Vinod Malik, Louise R. Rodino-Klapac, Brian K. Kaspar, Lei Chen, et al., "AAV1.NT3 Gene Therapy for Charcot-Marie-Tooth Neuropathy," *Molecular Therapy* 22, no. 3 (2014): 511–21, doi: 10.1038/mt.2013.250.

Conclusion

Epigraph: Oliver Sacks, *The River of Consciousness* (New York: Vintage, 2017), 150.

1. Javier Apfel and Cynthia Kenyon, "Regulation of Lifespan by Sensory Perception in *Caenorhabditis elegans*," *Nature* 402, no. 6763 (1999): 804–9, doi: 10.1038/45544; Cynthia J. Kenyon, "The Genetics of Ageing," *Nature* 464, no. 7288 (2010): 504–12, doi: 10.1038/nature08980.

2. Koutarou D. Kimuri, Heidi A. Tissenbaum, Yanxia Liu, and Gary Ruvkun, "Daf-2, an Insulin Receptor-Like Gene that Regulates Longevity and Diapause in *Caenorhabditis elegans*," *Science* 277, no. 5328 (1997): 942–46, doi: 10.1126/science.277.5328.942.

3. Joy Alcedo, Thomas Flatt, and Elena G. Pasyukova, "Neuronal Inputs and Outputs of Aging and Longevity," *Frontiers in Genetics* 4 (2013): 1–14, doi: 10.3389/fgene.2013.00071.

4. Coleen T. Murphy, Steven A. McCarroll, Cornelia I. Bargmann, Andrew Fraser, Ravi S. Kamath, Julie Ahringer, Hao Li, and Cynthia Kenyon, "Genes that Act Downstream of DAF-16 to Influence the Lifespan of *Caenorhabditis elegans*," *Nature* 424, no. 6946 (2003): 277–83, doi: 10.1038/nature01789.

5. Luigi Fontana, Linda Partridge, and Valter D. Longo, "Extending Healthy Life Span—From Yeast to Humans," *Science* 328, no. 5976 (2010): 321–26, doi: 10.1126/science.1172539.

6. Céline E. Riera, Mark O. Huising, Patricia Follett, Mathias Leblanc, Jonathan Halloran, Roger Van Andel, Carlos Daniel de Magalhaes Filho, Carsten Merkwirth, and Andrew

Dillin, "TRPV1 Pain Receptors Regulate Longevity and Metabolism by Neuropeptide Signaling," *Cell* 157, no. 5 (2014): 1023–36, doi: 10.1016/j.cell.2014.03.051.

7. Ibid.

8. Ibid.

9. Soumitra Ghosh, Lin Li, and Warren G. Tourtellotte, "Retrograde Nerve Growth Factor Signaling Abnormalities and the Pathogenesis of Familial Dysautonomia," *Neural Regeneration Research* 16, no. 9 (2021): 1795–96, doi: 10.4103/1673-5374.306081.

10. Rachel M. Easton, Thomas L. Deckworth, Alexander Sh. Parsadanian, and Eugene M. Johnson Jr., "Analysis of the Mechanism of Loss of Trophic Factor Dependence Associated with Neuronal Maturation: A Phenotype Indistinguishable from Bax Deletion," *Journal of Neuroscience* 17, no. 24 (1997): 9656–66, doi: 10.1523/JNEUROSCI.17-24-09656.1997.

11. Ibid.

ACKNOWLEDGMENTS

The journey through the peripheral nervous system has taken us to many topics and unlikely connections. I am lucky to be acquainted with many scientists who have the ability to raise major questions that have not been addressed. The rationale for *Periphery* actually came from a rare and unknown genetic disorder called familial dysautonomia (FD), or Riley-Day syndrome, a malady that strikes the whole body through the periphery. I am indebted to Felicia Axelrod, MD, and Horacio Kaufmann, MD, faculty members at New York University, who pioneered the study of FD. Their offices and laboratories lie two floors above my office at the Skirball Institute, illustrating the importance of proximity in scientific interactions. Distant connections with Professor Frances Lefcort (Montana State University) also stimulated my interest in the subject.

On a visit to the Queen Square Institute of Neurology at University College London (UCL), I became acquainted with a number of disorders, including ALS, Alzheimer's, and Huntington's diseases Giampietro Schiavo, John Hardy, and Sarah Tabrizi, who were expert consultants, kindly provided me with office space at UCL to initiate this project and directed me to source material. Others at Project ALS kept me abreast of exciting developments over the past decade that spill over to other diseases. I am appreciative of the many foundations that freely shared their research efforts and created more public and political support for basic research in human diseases.

On my visits to London, many friends and colleagues were sources of information regarding current concepts in neurobiology and neurology. They include Martin Raff, Sarah Caddick, Frank Walsh, Pat Doherty, John Wood, and Stephen McMahon. Wood and McMahon were indispensable resources regarding the subject of pain. I was deeply saddened to learn of Stephen McMahon's untimely death in 2021.

Historical accounts of autism research revealed unexpected relationships between enlightened physician-scientists such as James Parkinson, Jean-Martin Charcot, and Sigmund Freud. A more modern influence came from Gerry Fischbach, who directed SFARI, the Simons Foundation Autism Research Initiative. Many meetings and generous financial support from the Simons Foundation paved the way for the identification of autism genes and therapeutic approaches. In particular, David Ginty and Lauren Orefice (Harvard University) pioneered the idea that the causes of autism stemmed from alterations of sensory functions (sound, touch, taste, sight, hearing)—a simple yet elegant explanation for a complex disorder.

My laboratory and the Skirball Institute at NYU provided the optimal atmosphere for exchanging new and unconventional ideas. Breakthroughs in science do not arise from a straight-line approach but often require a spark of imagination from other fields. Since the Skirball Institute was organized around questions and not departments, there was a greater acceptance of new approaches in genetics and cell biology not taken by others. I credit my editor, Janice Audet, for her commitment and continuous interest in the subject; she was very effective in smoothing rough edges and anticipating next steps. Avraham Yaron provided clear diagrams, and Emeralde Jensen-Roberts and Stephanie Vyce made sure the images were processed and fit the occasion in an appropriate way.

CREDITS

2.1 Reprinted from Rui B. Chang, David E. Strochlic, Erika K. Williams, Benjamin D. Umans, and Stephen D. Liberies, "Vagal Sensory Neuron Subtypes that Differentially Control Breathing," *Cell* 161, no. 3 (2015), 622–33, figure 2A, by permission of Elsevier.

3.1 Courtesy of Avraham Yaron, Weizmann Institute.

4.1 Data source: Patrick Haggard and Sam Rogers, "The Movement Disorder of Nicholas Poussin (1594–1665)," *Movement Disorders* 15, no. 2 (2000): 328–34, figure 3.

5.1 W.Y. Sunshine/Shutterstock.

5.2 Reprinted from Betty Ben Geren, "The Formation from the Schwann Cell Surface of Myelin in the Peripheral Nerves of Chick Embryos," *Experimental Cell Research* 7 (1954): 558–62, figure 5, by permission of Elsevier.

5.3 Designua/Shutterstock.

5.4 Detail reprinted from J. L. Salzerand and B. Zalc, "Myelination," *Current Biology* 26 (2016): R937–80, figure 1, by permission of Elsevier.

5.5 Reprinted from John G. Nicholls, A. Robert Martin, Paul A. Fuchs, David A. Brown, Mathew E. Diamond, and David A. Weisblatt, *From Neuron to Brain*, 2011, figure 27.18. Used with permission of Oxford University Press conveyed through Copyright Clearance Center, Inc.

6.1 Courtesy of Avraham Yaron, Weizmann Institute.

7.1 Reprinted from Christina T. Fuentes, Stewart H. Mostofsky, and Amy J. Bastian, "Children with Autism Show Specific Handwriting Impairments," *Neurology* 73, no. 19 (2009): 1532–37, figure 1. With permission from Wolters Kluwer Health, Inc.

8.1 Chen I Chun/Shutterstock.

C.1 Data source: Joy Alcedo, Thomas Flatt, and Elena G. Pasyukova, "Neuronal Inputs and Outputs of Aging and Longevity," *Frontiers in Genetics* 4 (2013), figure 1.

INDEX